P9-DWZ-012

Who Got Einstein's Office?

Addison-Wesley Publishing Company, Inc.

Reading, Massachusetts
Menlo Park, California
New York
Don Mills, Ontario
Wokingham, England
Amsterdam
Bonn
Sidney
Singapore
Tokyo
Madrid
San Juan
Paris
Seoul
Milan
Mexico City
Taipei

Who Got Einstein's Office?

Eccentricity and Genius at the Institute for Advanced Study

Ed Regis

Many of the designations used by manufacturers and sellers to distinguish their products are claimed as trademarks. Where those designations appear in this book and Addison-Wesley was aware of a trademark claim, the designations have been printed in initial capital letters (i.e., Macintosh).

Library of Congress Cataloging-in-Publication Data

Regis, Edward, 1944–
Who got Einstein's office?

Bibliography: p.
Includes index.
1. Research institutes—New Jersey—Princeton—History. 2. Research——New Jersey—Princeton—History. 3. Institute for Advanced Study (Princeton, N.J.)—History. I. Title.
Q180.U5R35 1987 001.4'09749'67 87-11568
ISBN 0-201-12065-8
ISBN 0-201-12278-2 (pbk.)

Cover design by Skolos Wedell + Raynor, Inc.
Text design by Paul Souza/Perfect Design Sense
Set in 10-point Palatino by Publication Services, Champaign, IL

5 6 7 8 9 10 - DO - 949392

Fifth printing, April 1992

Excerpts from *Abraham Flexner: An Autobiography* (New York: Simon & Schuster, 1960), ©1940 by Abraham Flexner, ©1960 by Jean Flexner Lewison and Eleanor Flexner. Reprinted by permission of Simon & Schuster, Inc.

Excerpt from N. David Mermin, "Is the moon there when nobody looks? Reality and the quantum theory." *Physics Today*, April 1985, ©1985 by the American Institute of Physics. Reprinted by permission.

Presentation of Cantor's diagonal argument adapted from Douglas Hofstadter, *Gödel, Escher, Bach* (New York: Basic Books, Inc., 1979).

The Logo Tree Display in Chapter 4 is reprinted by permission of the publisher of *Nibble Mac* magazine, 45 Winthrop St., Concord, MA 01742. Copyright by MicroSPARC, Inc., 1986. All rights reserved.

The Mandelbrot set displays in Chapter 4, pages 69, 80, and 91 were created using Robert P. Munafo's program Super ***MANDELZOOM*** 1.0. ©1986 by Robert P. Munafo. Used by permission.

The Mandelbrot set displays on pages 71 and 76 were created using John Milnor's BASIC program for the Mandelbrot set (see Appendix). Program ©1986 by John Milnor. Used by permission.

Excerpts from interview with Benoit B. Mandelbrot, *Omni*, February 1984, ©1984 by Omni Publications International, Ltd. Reprinted by permission.

Presentation of John von Neumann's treatment of quantum mechanics adapted from Steve J. Heims, *John von Neumann and Norbert Wiener: From Mathematics to the Technologies of Life and Death* (Cambridge, Mass.: The MIT Press, 1980).

Excerpts from P. R. Halmos, "The Legend of John von Neumann," *American Mathematical Monthly* 80: 382–394 (1973), ©1973 by The Mathematical Association of America. Reprinted by permission.

Excerpts from Herman H. Goldstine, *The Computer from Pascal to von Neumann*. Copyright ©1972 by Princeton University Press.

Excerpts from Oral History interview of J. Robert Oppenheimer by Thomas S. Kuhn, November 18 and 20, 1963. Archive for History of Quantum Physics, copy at the American Institute of Physics, New York City. Reprinted by permission.

The account of the Project Orion tests on Point Loma was derived in part from Kenneth Brower, *The Starship and the Canoe* (New York: Harper and Row, 1983). ©1978 by Kenneth Brower.

Excerpts from Oral History interview of P.A.M. Dirac by Thomas S. Kuhn, January 4, 1962, and May 7, 1963. Archive for History of Quantum Physics, copy at the American Institute of Physics, New York City. Reprinted by permission.

Excerpts from John Bahcall's lecture "The Problem of Solar Neutrinos," February 26, 1976, at Vassar College, Poughkeepsie, New York. Unpublished draft transcript. Reprinted by permission.

The cellular automata displays in Chapter 10 were created using "Automata," a program written by Nicholas B. Tufillaro, Jeremiah P. Reilly, and Richard E. Crandall, ©1985 by the Reed Institute. Used by permission.

Some of the Life images in Chapter 10 were created using a public-domain program for John Conway's game of Life by Bill Atkinson, version 1.0.

Excerpts from Edward Regis Jr., "Einstein's Sanctum," *Omni*, September 1984, ©1984 by Edward Regis Jr. and used by permission of Omni Publications International, Ltd.

Excerpts from Paul Ginsparg and Sheldon Glashow, "Desperately Seeking Superstrings?" *Physics Today*, May 1986, ©1986 by the American Institute of Physics. Reprinted by permission.

Cellular automata program, Macintosh Microsoft BASIC version (see Appendix), copyright ©1986 by Nicholas B. Tufillaro. Used by permission.

Excerpt from *"Surely You're Joking, Mr. Feynman!" Adventures of a Curious Character*, by Richard P. Feynman, as told to Ralph Leighton. Edited by Edward Hutchings. Copyright ©1985 by Richard P. Feynman and Ralph Leighton. Used by permission.

To

Patrice Adcroft
Jane Bosveld
Doug Colligan
Dick Teresi
Pamela Weintraub
Gurney Williams III

Contents

Preface

I first came to the Institute for Advanced Study in the fall of 1983 to do a magazine story. Before I arrived on campus I had known the Institute only by reputation, as the place where Einstein and Gödel had worked for a large portion of their lives as scientists. As is probably true of other laymen interested in science, pictures of Einstein's old office at the Institute made a great impression on me in my youth. I had seen them in biographies of the man, and also in books on twentieth-century science. They're famous pictures, taken shortly after he died, in April 1955. They show a blackboard covered with equations, an empty chair turned sideways—perhaps in the exact position that Einstein left it when he rose from his desk for the last time—and bookshelves full of haphazardly arranged volumes. Most of all, though, I was impressed by the clutter on the man's desk: papers, journals, manuscripts, a bottle of fountain-pen ink, a pipe, a tobacco humidor . . . the effluvia of unfinished cosmic business. What undiscovered secrets of the universe, I wondered, lay hidden in the disorder?

Also in my memory was a photograph of another scientist. It had been taken in the Institute's mathematics library, and it pictured a gaunt man with a black stripe of hair running down an otherwise gray head, giving him a vaguely Mohawk look. The effect was repeated by the expression on his face: he was staring daggers at the camera, as if he wished that it—and the photographer—would go back to where they came from. This was Kurt Gödel.

To me, Einstein and Gödel were the number-one and -two geniuses of contemporary science, and for both of them to be at the same place at the same time, in Princeton, New Jersey, was more than a little mysterious. How did it happen that the two of them ended up at the Institute for Advanced Study? What *was* the Institute, anyway, and what did its great minds actually *do* there? And what had happened to the place since Einstein and Gödel died?

I was convinced, at any rate, that the Institute for Advanced Study must be a very special place, and so indeed it is. Virtually all the great figures of twentieth-century physics and mathematics have been here at one time or another, including fourteen Nobel prizewinners, people such as Niels Bohr, P.A.M. Dirac, Wolfgang Pauli, I.I. Rabi, Murray Gell-Mann, C.N. ("Frank") Yang, and T.D. Lee. In 1980, the Institute published a book called *A Community of Scholars*, which is a listing of all of the various researchers who came to the Institute during the first fifty years of its existence. It's a big book, over 500 pages long, and it's not an easy task to think of a twentieth-century scientist whose name you could not find somewhere within.

Humanists have come to the Institute, too, but in far smaller numbers than scientists, and their names are not as illustrious, although there is an exception in the person of T.S. Eliot. Eliot aside, the Institute has not supported studies in literature or criticism. Rather it focuses on social science and history, disciplines in which progress is tenuous and elusive, at least as compared with the advances in science that have taken place over the last fifty years, which is also about the same length of time that the Institute has been in existence. The scientists who have come to the Institute for Advanced Study are the ones who have revolutionized physics and have brought us close to what is perhaps the complete and final theory of nature. They have taken us from the dawn of quantum mechanics to the verge of the grand unification—the Theory of Everything—in the span of a single human lifetime. The story of the Institute is the story of its scientists, and that is the story told in this book.

The Institute's scientists are not, on the whole, a modest lot. This is quite understandable. Their goal, after all, is the largest and most difficult that any group of people could possibly have set for themselves. They merely want to understand . . . *everything*, to know and explain the whole of nature. They want to comprehend why the physical universe is the way it is and why it works the way it does. The Institute exists to honor the arrogance of mind required of those who are immodest enough to think they can contribute to the task, and this book is my attempt to portray the lives and work of a few who have done so.

Ed Regis
Eldersburg, Maryland
December 15, 1986

Acknowledgments

Many people who are now or have been connected with the Institute for Advanced Study have helped me with this project, and I am indebted to them all for their time and effort. Portions of the manuscript were read by several past and present Institute members, as well as by others, and for their comments and corrections I owe thanks to Stephen Adler, John Bahcall, John Dawson, Rick Dillman, Freeman Dyson, Margaret Geller, Herman Goldstine, Jeremy Goodman, Charles Griswold, Banesh Hoffmann, Douglas Hofstadter, Andrew Lenard, Benoit Mandelbrot, N. David Mermin, John Milnor, Tim Morris, Mark Mueller, Abraham Pais, Harry Rosenzweig, Don Schneider, Dudley Shapere, and Stephen Wolfram. For the errors of fact or interpretation that may remain, I am alone responsible.

For personal favors and other assistance, I would like to thank Mr. and Mrs. Robert Bacher, Julian Bigelow, Jack Clark, Linda Eshleman, Joan Feast, Diana Howie, Priscilla Johnson McMillan, Robert P. Munafo, Keith Richwine, Paul Schuchman, Linda Sheldon, Nick Tufillaro, Caroline Underwood, Sterling White, Mary Wisnovsky, and most especially Flora Dean, who, as trustee for Beatrice M. Stern, provided me with a copy of the history of the Institute mentioned in the narrative.

For their hospitality in Princeton, I would like to thank Brock, Ann, and Alison Brower. I owe a special debt to my wife, Pamela Regis, for her research help, editorial advice, and for preparing the index.

Particular thanks go to Robert Lavelle, who, as acquisitions editor for Addison-Wesley Publishing Company, suggested that my *Omni* article on the Institute be turned into a book, and to senior editor William Patrick, for his good judgment and his knack for calming the author's sometimes frazzled nerves.

Since this project would never have seen the light of day were it not for my editors at *Omni*, I have dedicated the book to them, in thanks for their support over the years.

Prologue

Chapter 1

The Platonic Heaven

Princeton, New Jersey, was for many years a quiet prerevolutionary village known mainly for the Battle of Princeton, in which Washington and his men whipped the British, and for its university. Settled in 1685 by Quakers who were drawn to the area's flatlands, streams, and woods, Princeton was for six months—in 1783 when the Second Continental Congress met there—the capital of the United States. Before that, in 1756, Princeton became the home of the College of New Jersey, which had been founded by Presbyterians at the height of the Great Awakening, a frenzied revival of orthodox Calvinism. After a fund drive the college erected Nassau Hall—for a time the largest building in the Colonies—and invited the fire-and-brimstone preacher, Jonathan Edwards, to install himself there as president.

Edwards was the Connecticut theologian who, in the beloved and time-tested manner of the religious Platonists from Bishop Berkeley on, taught the doctrine of philosophical idealism, the belief that the external world is nothing but . . . *an idea*. "The World, i.e., the material Universe, exists no where but in the mind," he said, anticipating by some 200 years the Princeton scientists who would in their own way reduce the "material Universe" to a network of mental abstractions.

Edwards also preached the puzzling Calvinist dogma which said that, although God had decided long before your birth whether you're going to heaven or hell, you can nevertheless somehow—he was not too clear about this—choose for yourself where you're going to end up. Evidently, God had decided that Edwards should not be president of the College of New Jersey, for shortly after being sworn into that office he contracted smallpox and died. Much later, in 1896, the college was renamed Princeton University, but it wasn't until 1902, when Woodrow Wilson became its president, that the school was for the first time headed up by a layman.

In October 1933, virtually overnight, Princeton was transformed from a gentleman's college town into a world center for physics. Albert Einstein had arrived at the Institute for Advanced Study.

The Institute was to be a new kind of research center. It would have no students, no teachers, and no classes. The world's greatest scientists would gather there to do research, but they would have no laboratories, no machines, no apparatus for doing experiments. This was very much by design and intention. From the start, the Institute for Advanced Study was to be a hotbed of pure theory, something that Jonathan Edwards would certainly have approved of, for Edwards had gotten smallpox, not from other people, but from an anti-smallpox vaccine that he had agreed to test out. The practice of vaccination was then in its experimental stages, and Edwards wanted to demonstrate his faith in the wonders of modern science. So he volunteered himself for an inoculation, got smallpox, and died.

Today the Institute for Advanced Study is as much devoted to theory as it was at the beginning, although the absence of an Einstein or a Gödel on the faculty gives it a somewhat lower profile than it had back then. It's located right on the edge of town, but people who have lived in Princeton for their whole life often can't tell you where the Institute is or how to get there. You ask people at the university, a few blocks away, how to find the Institute for Advanced Study and they tell you they never heard of the place. "The Institute for *what*?" They can give you directions to the Princeton Theological Seminary, or to the Springdale Golf Club, but not to the Institute. As Homer Thompson, a faculty member there for the last forty years, says, "The Institute is better known in Europe than it is in Princeton."

People can hardly be blamed for their ignorance, for the Institute is one of those places you can't quite pigeonhole from the outside. Located on a square mile of open field and woodland on Olden Lane south of Princeton, it could be a college campus or a prep school, but there are never any students milling about, so that can't be what it is. And so you may begin to wonder. Is it a sanatorium? An orphanage? An old soldiers' home?

The main building, Fuld Hall, is a red-brick, Georgian-style structure such as you find on college campuses everywhere. Fuld Hall contains faculty and administrative offices, the mathematics library, and the common room, where tea and cookies are served every weekday at 3 o'clock. There are a few other, smaller buildings on either side of Fuld Hall, done in the same collegiate style, and which also contain offices. But then off to one side there's a complex of glass-and-concrete buildings that breaks the traditional pattern. They house the Institute's dining hall, social science offices, and historical studies library. Back of the library there's a small lake, and behind that, the woods.

When he first came to the Institute, in the 1940s, Freeman Dyson and a bunch of his friends used to ride through the woods in an old Dodge convertible with the top down. Nowadays, the only traffic through the forest consists of walkers, joggers, and deer hunters. The deer have gotten so bad that Allen Rowe, one of the associate directors, now sends around an annual memo explaining the Institute's "Deer Control Program." "To reduce the number of deer in the Institute woods to a reasonable and sustainable level," the memo says, the herd will be "culled out" by a small band of expert bow hunters. Every time the memo comes out, some of the younger members wonder if they should stage some kind of protest, but nothing has ever come of this. They just keep out of the woods for a few weeks. It's not proper Institute style to come back from a hike with an arrow through your head.

J. Robert Oppenheimer, who was director of the place for almost twenty years, used to describe the Institute as "an intellectual hotel," a refuge to which scholars could repair for any length of time and have their worldly needs taken care of by others. The usual length of stay is one or two years; and, in fact, of the 200 or so people who are in residence there at any one time, most are younger scholars on short-term memberships. The Institute is divided up into four schools, Mathematics, Natural Sciences, Historical Studies, and Social Science. Most of the members are in math and the physical sciences; the smallest school, Social Science, has only some twenty or so resident scholars per year.

To be admitted to this fine circle of friends, you must run the gauntlet and survive it, a winnowing-out procedure that leaves only the most extraterrestrially brilliant people still standing. If you should be lucky enough to have made it through, you'll receive a stipend, an office, and your own apartment in the housing project, a collection of faintly Bauhaus buildings designed by architect Marcel Breuer. From the time of your arrival until the day of your departure, the Institute will offer you breakfast and lunch five days a week, and dinner on Wednesday and Friday nights. There is never a complaint about any of this. Rarely, it's true, you hear some muttering about the furniture. "The chairs in the apartments were terribly uncomfortable," recalls one member. "The desk chair was fine. One almost got the impression that there was a conspiracy to encourage you to work every single minute."

And then there's the matter of the beds. Not that they're uncomfy, it's just that they're all single beds, not a double bed in the place . . . except, perhaps, for Olden Manor, where the director and his wife live. The younger couples find this a bit amusing. "They either got all these beds at a really great fire sale," says one member, "or they just want you to keep your mind on your work."

Other than for the sleeping arrangements, the Institute is exactly the nirvana for eggheads that it's cracked up to be. The temporary members, and the far smaller group of permanent professors—there are about twenty-five of these—work on their own projects at their own pace and have no further responsibility or accountability to anyone. At the end of your stay you don't even have to write up a report describing the work you did there, if any; you just pack up and leave your intellectual Garden of Eden and return to the cruel world from which you came.

As perhaps befits a modern utopia, the Institute's professors, who are appointed on a permanent basis, are all paid at the same rate, now in the neighborhood of some $90,000 per annum, a figure that invites outsiders to speak of the place as the "Institute for Advanced Salaries." The Institute has an operating budget of over $10 million per year, which it pays for out of its endowment, currently valued at well over $100 million, and out of the income derived from its investments. The rate of return on its investments varies from year to year, but recently it has averaged about 17 percent per annum. In fiscal year 1984/5, though, the annual rate of return was 26.9 percent. Unlike other utopias envisioned by idealists down through the ages, this one is not afraid of money.

The Institute for Advanced Study was founded in 1930 by Louis Bamberger and his sister, Caroline Bamberger Fuld—they put up the cash—and by Abraham Flexner, who planned and organized it all. But the Institute's true father, intellectual progenitor, and watchful guiding spirit is the ancient Greek philosopher, Plato. For one thing, it was Plato who long ago established the world's first institution of higher learning—called the Academy—located on the outskirts of Athens. Here scholars, researchers, and theorists of every stripe came together to fathom the plan of the world, to try to comprehend under a single intellectual edifice the overall scheme of things. Plato's was the first large-scale, systematic effort to poetize the phenomena, to reduce the whole visible universe to a small set of abstract concepts and principles. The Institute for Advanced Study is right in the Platonic mold, but this only begins to tell why Plato is its true father.

We're getting closer when we realize that for Plato the true objects of knowledge are not the transient and changing entities that we can see and touch, but something else entirely, something that Plato thought was ultimately more real, things he called the "Forms." Plato looked around at nature, looked at the sun and the moon and the stars, he looked at waterfalls, plants, and animals, and summarily declared that these gross physical objects were not the True Reality at all. True Reality lay in another dimension altogether, one he styled as the World of the Forms. The World of the Forms is invisible to the senses, and for this reason one might be tempted to regard it as a shadowy, murky never-never land, but Plato himself never thought of it that way. To him it was as bright and luminous as the noonday sun. The World of the Forms, after all, was where all this world's physical objects came from. It was the Origin, the Source of everything.

For Plato, the major difference between Forms and the ordinary objects of everyday life is that, whereas ordinary objects are changing and corruptible, the Forms are perfect, unchanging, and eternal. Because they are immutable, the Forms are more real than "real" things, but because they're in their own separate realm, you can't see the Forms or apprehend them by means of the senses. That, after all, would be much too easy. To come to know the Forms, you've got to do the hard thing: you've got to close your eyes, withdraw inward, and *think*. You've got to do an awful lot of this turning inward and thinking, in fact, which of course explains why the man on the street knows little or nothing about the Forms.

The scientists at the Institute for Advanced Study, to be sure, don't believe in any literal version of Plato's Forms, but they do study objects that go profoundly beyond sense experience, and which are apprehensible only by means of thought. The Institute's mathematicians, for example, rarely if ever bother with the tangible stuff of the world. On the contrary, they study abstract and idealized mathematical objects, entities that don't exist anywhere in nature. You will not find a circle anywhere in the real world, although there are plenty of things out there that are at least approximately circular. To a mathematician, an abstract geometrical circle is much more "real" than any nearly circular object is. The wheel on your car changes its shape from moment to moment and loses its rubber constantly, but a mathematical circle is perfectly round always and forever.

But it's not only the Institute's mathematicians who live and work in the midst of impalpable entities, for so do its physicists and astronomers. No object on terra firma lasts for very long—even mountain ranges wear down over mere millions of years—but the lifetimes of stars and galaxies are measured in the billions. Nature's elementary particles, protons and electrons, for example, have life spans that are virtually infinite, making these unseen entities more like Plato's Forms than they are like anything else.

Which is all another way of saying that the scientists who work at the Institute for Advanced Study don't get down in the mud and root around in the dirt of nature. Geologists, biologists, and open-heart surgeons need not apply here: such people have dirty hands and will not be admitted into this band of pure and proud theorists. The Institute's scientists gather at the roof of nature and contemplate impalpable objects at the farthest extremes of creation. They make no product and they do no experiment. Their whole purpose in life is simply and solely to *understand*.

"I work in celestial mechanics," Institute mathematician Marston Morse once said, "but I am not interested in getting to the moon."

For a place with such ethereal pretensions, the Institute was established on foundations that could hardly be more worldly, based as it was on the ring of cash registers and the clink of coins. The Institute for Advanced Study owes its existence to the New Jersey department store called Bamberger's, which in 1929 was doing $35 million worth of business off of one million square feet of floor space, making it the fourth largest retail store in the country.

Fortunately for the world of theoretical science, Louis Bamberger and his sister Caroline Bamberger Fuld, the store's owners, sold out of the business in 1929, just before the stock market crash. They contracted with the R.H. Macy Company to transfer ownership of the store in return for a combination of cash and Macy stock equal in value to $25 million. Bamberger received his money early in September, about six weeks before Black Thursday. Splitting the proceeds evenly with his sister, Bamberger distributed one million dollars of his own money to his former executives.

That gesture was just the beginning of a gigantic money giveaway. Bamberger and Fuld were of a genuinely philanthropic nature and felt that, since their money came from the good people of New Jersey who shopped every day at their store, it was only right to reward them for it; and so the two of them decided to endow a large public institution of some kind, one that would benefit all New Jerseyans. They had in mind a dental or medical college, which they were thinking of putting on their South Orange estate, but the problem was that neither of them knew the first thing about medical colleges other than that medicine was taught there. Soon, however, they found out the name of the one man who probably knew more about medical colleges than anyone else in the world. That man was Abraham Flexner.

Flexner was the hanging judge of American higher education. His one great mission in life was to improve the American college system, and his main claim to fame was "the Flexner Report," an exposé of quackery and fraud in the nation's medical schools. Later, as a foundation administrator, Flexner was personally responsible for prying from the wallets of American philanthropists more than a half billion dollars—$600 million, to be exact—which he then proceeded to dish out to the colleges that would, in his opinion, make the best use of it.

In money matters, as in most things, Flexner was a man of unwavering honesty; he had come up the hard way, and he knew the virtue of adhering to principle as well as he did the value of a dollar. Born in Louisville, Kentucky, in 1866, Abe was the sixth of nine children raised by Moritz and Esther Flexner, who had migrated to the United States in the middle of the nineteenth century. Moritz, starting out as a peddler, later managed to gain control of a wholesale hat business, which he nurtured for many years only to see it wiped out in the Great Panic of 1873.

In those days, education in the South ranged from poor to nonexistent, and young Abe was enrolled in a typically backward elementary school. Most of the time he had no homework to do, and so he would go to the library and read books on his own. Somehow he latched onto the classics: Dickens, Shakespeare, Thoreau, Hawthorne. Later, as a teenager, he took a job at a private library, which gave him plenty of time to read and to listen to the small discussion groups that met there once a week to talk over current events. So far, Abe was self-educated. All that changed, however, when he went away to college, to Johns Hopkins.

"The decisive moment of my life," Flexner said later, "came in 1884 when, at the age of seventeen, I was sent by my oldest brother Jacob, to the Johns Hopkins University. Upon that choice my whole subsequent career depended."

Hopkins had just opened its doors, and Abe watched it grow from the ground up, from its beginnings in two converted boarding houses on Baltimore's Howard Street, to its later home on a munificent parklike campus in the center of the city. Hopkins became the model upon which Flexner would later build the Institute for Advanced Study.

The Johns Hopkins University was named after the wealthy Baltimore merchant who, when he died in 1873, left practically his whole fortune, $7 million—at that time the largest bequest ever made in the United States—for the purpose of setting up an institution which was to be half university, half hospital. The then extant colleges, Harvard, Yale, Columbia, and others, were offering only fair-to-middling educations, nothing like the advanced graduate programs of today, and so when Hopkins opened in 1876 it was the first time America saw a graduate school in the modern sense of the term.

In Flexner's mind, Johns Hopkins was all that was good, right, and true in higher education, and the university's first president, Daniel Coit Gilman, was Flexner's idol. To start the college, Gilman had scoured Europe for first-rate faculty members, but he wasn't looking only for chalk-and-eraser men, i.e., classroom teachers. He also wanted creative people, and good researchers. "He knew that there are men who teach best by not teaching at all," Flexner said of Gilman. *Teach best by not teaching at all.* The idea was that the true geniuses, people like Darwin, Faraday, Rayleigh, and so on, had made their mark on the world precisely because they didn't have to spend their valuable time on such things as preparing for and teaching classes. Flexner never forgot this point.

When Flexner graduated from Hopkins he returned to Kentucky to teach, ending up as a teacher at the same high school he had attended as a student only a few years earlier. He had gone through Hopkins in just two years and so, when he got back to his old high school, he found he was teaching some of the same youngsters who had once been his classmates. At the end of his first year, Flexner flunked an entire classful of students—all eleven of them. This unprecedented act stunned the parents of Louisville and made the local newspapers. The parents demanded that their good and lovely children be promoted forthwith, and they called for a public hearing before the school board. The board met, listened to the evidence, and promptly sustained Flexner, thereby making his local reputation as a no-nonsense educator who upheld his standards come what may.

Thereafter, people who wanted their sons and daughters to learn more than the local schools were teaching them started bringing their kids to Flexner. An attorney whose son had been thrown out of prep school asked Flexner to tutor the boy and get him into Princeton. Flexner agreed, but only on the condition that the lawyer also bring him four other students to be tutored at the same rate, which was $500 per year. This the lawyer did, and soon enough Flexner quit his high school job and set up his own little prep school.

"Mr. Flexner's School," as people called it, was a phenomenal success. Flexner took on any student no matter how dull or recalcitrant, at the same time promising parents he'd whip their kids into shape and get them into Princeton, Harvard, or anywhere else. The astonishing thing was, Flexner succeeded, and he did so without threats, force, or pressure. "I had long since learned that I could accomplish nothing by compulsion," he said later. "The school operated without rules, without examinations, without records, and without reports."

Much like faculty members at the Institute for Advanced Study later on, the students at Mr. Flexner's school had no responsibilities. "No duties, only opportunities," that was always Flexner's attitude. His students could come to class or stay away. They could drop in when they wanted and do as much or as little as they wished. Perhaps even to his own surprise, Flexner's kids started dropping by on Saturdays, just to get in a little extra work. There was no miracle involved: Flexner did it all by sheer force of personality and a genuine enthusiasm for learning.

Flexner's home-grown prep school lasted for fifteen years, from 1890 to 1905, when the master decided to go back to school himself. He enrolled in Harvard for a psychology degree, but found that the required experimental work was so boring that he couldn't continue, and so he dropped out and went back to the process of self-education. This time, though, he'd take as his subject not some dull and narrow academic discipline like experimental psychology but the entire institution of higher learning in the United States. The end product was a scorching critique of the whole system, his book, *The American College*.

In it, Flexner claimed that colleges stifled exactly the capacity they were supposed to develop, individual initiative on the part of the student. He complained that colleges placed a higher value on specialized research than on good teaching, and that in many cases undergraduate curriculum simply made no sense. Not many educators wanted to hear his message, of course, but Flexner did have a few listeners. One of them was the Carnegie Foundation for the Advancement of Teaching, which hired Flexner to make a special study of medical schools.

At the turn of the century, American medical "colleges" generally operated in the manner of today's institutes of high-fashion modeling, computer repair, or advanced semitrailer trucking. That is to say, they took your money, gave you a class or two, and handed you an enormous diploma with your name on it in extremely large Old English script. These bogus medical schools sprang up everywhere, like weeds. There were forty-two of them in the state of Missouri, and fourteen in the city of Chicago alone. Classes, of course, were optional. You paid your tuition and a year later you were a doctor, no muss, no fuss.

Flexner gave himself a crash course in medical education, visiting the School of Medicine at his beloved Johns Hopkins, and also the Rockefeller Institute for Medical Research, in New York City, where his brother, Simon Flexner, was director of laboratories. Using their programs as paradigms of what might and ought to be, Flexner took off to see the sights at medical institutions all across the country. To ferret out the truth, he often resorted to trickery.

"On [one] occasion," he wrote, "I visited one morning an osteopathic school in Des Moines with the dean of the school and found every door locked, though on the outside each bore a name—ANATOMY, PHYSIOLOGY, PATHOLOGY, etc. The janitor could not be found—probably not altogether an accident. Having expressed my satisfaction, I was driven to the railroad station and left, as the dean supposed, to take the next train to Iowa City. Instead, after waiting until he got out of sight, I returned to the school, found the janitor, gave him five dollars, in return for which he opened every door. The equipment in every one of the rooms was identical. It consisted of a desk, a small blackboard, and chairs; there were no charts, no apparatus—nothing!"

Flexner personally inspected every last one of the 155 medical colleges in the United States and Canada. He discovered that only some half dozen of them maintained proper standards of admission, teaching, and graduation. Flexner had choice words for the rest: "Nothing more disgraceful calling itself a medical school can be found anywhere," he said of the Georgia College of Eclectic Medicine and Surgery. The California Medical College, likewise, was "a disgrace to the state whose laws permit its existence."

Flexner submitted a report that was first published as *Bulletin No. 4* of the Carnegie Foundation for the Advancement of Teaching. Later on it became known simply as "The Flexner Report." The book was an even bigger sensation than *The American College*, garnering for its author not only a succession of libel suits but also a threat on his life. An anonymous letter from Chicago—"the plague spot of the country in respect of medical education," he'd called it—said he would be shot if he ever showed up again in that fair city. "Whereupon," he said later, "I went there to make a speech before a meeting called by the Council on Medical Education and returned unharmed."

But Flexner had won his battle. Quack medical colleges folded up and faded away, often without a peep. Chicago's fourteen schools were winnowed down to three. In educational circles Flexner became a celebrity, looked upon as the knight savior of American medicine. Soon he was doing the same study in Europe, sent there by the Carnegie Foundation. By the time he was through with it all, Abraham Flexner was the world's number-one expert on medical education.

In the fall of 1929, Louis Bamberger and his sister Caroline (Mrs. Felix) Fuld were thinking about endowing a medical college in New Jersey. At the same time Abraham Flexner was writing a new book, *Universities: American, English, German,* based upon a series of lectures he had given at Oxford. "I was working quietly one day," he remembered, "when the telephone rang and I was asked to see two gentlemen who wished to discuss with me the possible uses to which a considerable sum of money might be placed." The two men, Samuel Leidesdorf and Herbert Maass, represented Louis Bamberger and Mrs. Felix Fuld, who needed his help in starting a medical school in Newark. The college was to give preferential treatment to Jews, for Bamberger and Mrs. Fuld, who were Jewish, were convinced that existing medical institutions discriminated against Jews in both staff and students.

Flexner, who was Jewish himself, wasn't buying any of this. For one thing, he said, a decent medical school would have to be attached to a major university, as well as to a good hospital, neither of which existed in Newark. Moreover, Newark was right across the river from New York City, which had excellent schools that they could not possibly hope to compete with. Finally, Flexner was convinced, on the basis of his extensive personal experience, that medical colleges did not in fact discriminate against Jews and that, in any case, he could not affiliate himself with an institution that used anything but the highest professional standards in the selection of either students or personnel.

On the other hand, these two men represented a fortune of some $30 million. Flexner did not want to turn them away, so he proposed an alternative idea. "Have you ever dreamed a dream?" he asked them. Flexner certainly had, and there it was, all written down, in the manuscript of his new book, the first chapter of which was lying on his desk at that very moment.

But contrary to expectation, Flexner was not dreaming of a research institute. In principle, he was against pure research institutes and had been for some time. In November of 1922, in fact, he had written down his objections to them in a report he had prepared for the General Education Board, a Rockefeller educational foundation. The report was entitled "A Proposal to Establish an American University" and in it Flexner stated, "Research institutions, valuable and necessary as they are, cannot alone remedy the difficulty [that graduate schools are in]—first, because relatively few men are most happy and effective if their entire energies are concentrated solely upon research; second, because the number of young men who can be trained in research institutions is necessarily limited. . . . Research institutions cannot . . . take the place of universities where men receive higher training."

Nevertheless, Flexner was not really satisfied with graduate schools as they then existed. Either the students who came out of them were poorly trained because faculty members spent all their time doing research, or they came out competent only for some specific practical vocation, such as law or medicine. In no case, so far as Flexner was concerned, did the students and faculty of any institution mutually cooperate in forming a true "society of scholars," by which he meant one whose goal was simply and solely to advance the frontiers of knowledge, to explore the unknown. But this was exactly Flexner's vision. He dreamt of a hybrid university—or an institute of some type, what you called it didn't matter—in which faculty members and students went into virgin territories *together*, not as equals, perhaps, but at least as partners. What you needed for this, he thought, was maximum freedom from outside pressure.

"It should be a free society of scholars," he wrote. "Free, because mature persons, animated by intellectual purposes, must be left to pursue their own ends in their own way." It should furnish simple surroundings, "and above all tranquillity—absence of distraction either by worldly concerns or by parental responsibility for an immature student body."

Flexner was writing all this down in the book he was working on, and he gave his callers, Leidesdorf and Maass, a copy of the first chapter, which he had entitled "The Idea of a Modern University." They took it back to Bamberger and Mrs. Fuld, who, after reading it, entertained no more plans for a medical school.

During the fall concert season, Bamberger and Fuld kept a suite in the Hotel Madison, in New York, and one evening they invited Flexner there for dinner. In succeeding weeks he met them regularly for lunch, and by mid-January, just before they were to leave for their winter vacation in Arizona, Flexner wrote up a set of working papers outlining how their fortune could best be utilized. The papers described "the endowment of an institution of higher learning situated in or near the City of Newark and called after the State of New Jersey in grateful recognition of the opportunities which we have enjoyed in that community." By the time Bamberger and Mrs. Fuld were setting off for the sunny Southwest, all three parties understood that they would soon be founding the university of Flexner's dreams.

When the two returned from Arizona in April, the only thing that had changed was the name of the place. Now it would be "an Institute of Higher Learning or Advanced Studies." It would have no undergraduate students; its faculty members would be well paid; and, above all, both students and faculty would dedicate themselves to new and fundamental explorations.

The Institute was formally and officially incorporated on May 20, 1930, as the Institute for Advanced Study; it opened its doors three years later. In between, there were two problems to be ironed out. One was location. Bamberger still longed to put the Institute in South Orange, but failing that he insisted anyhow it must be located "in the vicinity of Newark." The trouble was, Newark may have been a prime location for a dry-goods business, but it was about the last place on earth you'd want to put such a great new venture in higher learning. Nothing about Newark, Flexner thought, would attract anyone of an academic bent: the city had no university, no great libraries, museums, or collections. Mainly, it had a lot of paint and varnish factories. At one point Flexner sent a batch of letters to some forty educators around the country asking for their advice about where to put the Institute. They sent him a list of cities he could have predicted himself—New York, Cambridge, Chicago, Philadelphia, and so on. Not one of them mentioned Newark.

The most imaginative suggestion Flexner ever got came from Princeton University mathematician Solomon Lefschetz. Mindful of Bamberger's intention to put the Institute *somewhere* in New Jersey, Lefschetz argued that by act of creative reasoning the Institute could be placed in Washington, D.C. His idea was that since the nation's capital belonged to all of the forty-eight states it could by extension be considered as part of one of them—New Jersey. This type of desperation even began to infect Flexner, who began to wonder if the phrase "in the vicinity of Newark" could be stretched to encompass all of the northern, and perhaps even the central, part of the state.

Flexner, anyhow, had long since decided that the Institute would go up in Princeton, which he considered the ideal location. Princeton was removed from big-city distractions and pressures, yet within striking distance of New York, Philadelphia, and Washington. Its university already possessed one of the world's great mathematics departments, and it had a good library, to which Institute members would have visiting privileges. Flexner was going through the motions of searching for a site near Newark, but he was in contact with a real estate agent in Princeton.

The other problem was faculty. The Institute would sink or swim, Flexner thought, on the basis of its opening staff, and for this reason he wanted the best people he could find anywhere. But never in any dreams he dreamt did he imagine that the Institute's first professor would be the radiant light of the firmament, Albert Einstein.

The Priesthood of the Cosmos

Chapter 2

The Pope of Physics

"You're writing a book on the Institute . . . so maybe you can tell me," says Rob Tubbs. Tubbs is one of the Institute's short-term members, a young mathematician specializing in transcendental number theory. We're leaving his office after an interview session, and he pulls the door closed and locks it.

"A lot of us have heard these rumors about how they left Einstein's office just as it was the day he died, that they haven't touched a thing. So, uh . . . is that really true, or what?"

Well, why not ask the question? It's the same thing everyone else assumes when they first come to the Institute. This is where Einstein was, for more than twenty years. . . . *Einstein,* the greatest scientist who ever lived. . . . *Einstein,* the one and only scientist whose name absolutely anyone can produce on short notice. Why wouldn't his office be preserved? . . . as even the man's very brain was, which is now floating in a jar of formaldehyde in the office of one Thomas Harvey, M.D., of Weston, Missouri. Surely they must have closed up Einstein's office, maybe sealed it off forever, like a time capsule, otherwise it would be . . . a profanation, a desecration, a sacrilege. Who could possibly work there? Who could fill his shoes? Who could sit in that same office, day in, day out, year after year, knowing that *here is where Albert Einstein did his thing?*

"Where is his office, anyway?" Rob Tubbs asks.

Albert Einstein was a world cult figure long before he came to the Institute for Advanced Study. When in 1919 astronomers confirmed his prediction that light rays would be bent by the sun's gravity, people went crazy. They named babies and cigars after him. The London Palladium offered him a three-week stand, asking him to name his own price. Two German professors made a "relativity film" and showed it on both sides of the Atlantic. When Einstein entered the home of J.B.S. Haldane to stay the night, Haldane's daughter took one look at the man and promptly fainted dead away. The press hailed Einstein's theories as the greatest achievement in the history of human thought, and Einstein himself as the greatest man who ever lived.

He was, after all, the messenger of the new order. Light has weight, space is warped, the universe has four dimensions. People loved it. They didn't know what any of it meant, of course, but that didn't matter. Einstein did. He was the man who invented it all, who understood it. He became their hero, the new messiah, the First Knowledgian and Supreme Head of the Vast Physical Universe.

Einstein was revered as a god, but the man himself was the essence of modesty and kindliness, and he never understood why people made such a fuss. He, at least, treated others democratically, as equals: "I speak to everyone in the same way," he said, "whether he is the garbage man or the President of the University." Of course, if you were singleminded about it you could find some . . . exceptions, as for example the time he sent a paper to the *Physical Review*, and the editor dared to return it for revisions. *Well!* The poor editor had only done his job, sending Einstein's article—the same as he would anyone else's—to outside referees for evaluation. But this was not acceptable to Albert Einstein, who never sent one of his articles *there* again. But what does this prove? Only that the greatest physicist the world has ever known had, after all, an ego. If there's anything that most of the Institute's so-called prima donnas of science share, it's a healthy and well-developed ego.

In the world of ordinary men, Einstein may have been the humble genius who never wore socks (at least he wore shoes), but up there in the Platonic Heaven it was something else again. The man had this absolutely incredible hubris. He thought he might be able to understand the plan of the entire universe—the whole thing, from the largest galaxy to the smallest quark. He thought he could comprehend it all, that he could find a single overarching set of principles that would cover *everything*, in a theory of the unified field. And how would he do this but by theorizing, in the best Platonic-heavenly tradition. While the cyclotron people smashed their atoms to kingdom come, while the astronomers aimed their gigantic telescopes across billions of frigid light years, Einstein would close himself off in a room, pull the shades down, and, as he used to say, "I will a little think." He'd scribble out a few equations, make a few mental jottings, and lo and behold pretty soon he'd figure it all out. Just by thinking . . . no machines or instruments for him.

Someone once asked the great physicist where he kept his laboratory. Einstein smiled and took a fountain pen out of his breast pocket. "Here," he said.

In the winter of 1932 Abraham Flexner was off in California looking for faculty members for his new Institute. A Caltech professor by the name of Morgan suggested that Flexner pay a call on Einstein, who just happened to be there in residence. Einstein liked the idea of the Institute right from the start. Things were going from bad to worse back in Germany, where Einstein was a professor at the University of Berlin. In 1920 an anti-Einstein club had sprung up. Calling themselves the "Study Group of German Natural Philosophers," they offered money to anyone who would speak out against "Jewish physics," especially all this "relativity" business. On August 24, 1920, the "Antirelativity Theory Company Ltd.," as Einstein called them, sponsored a meeting in Berlin's Philharmonic Hall, which Einstein attended. He had to laugh, their attacks were that absurd.

These Aryan physicists were quite serious, however, and Einstein had endured their attentions for ten years. By 1931—in which year the "Antirelativity Company" published a book called *100 Authors Against Einstein*—he had decided to leave Germany for good, and so he was listening quite closely to Abraham Flexner as the two of them paced up and down the hallways of Caltech's faculty club discussing the new research center to be built in Princeton. Einstein asked to see Flexner again, and so they agreed to meet during the spring semester of 1932, when both of them would be in Oxford, England.

It was a beautiful Saturday morning in May. The sky was clear and the birds were singing and Flexner and Einstein were strolling back and forth across the Christ Church lawn like two proper Oxford dons when Flexner decided to pop the question. "Professor Einstein," he said, "I would not presume to offer you a post in this new institute, but if on reflection you decide that it would afford you the opportunities which you value, you would be welcome on your own terms."

Einstein was tempted, but all the same he wasn't about to be rushed into anything. Universities all over the place, including those at Jerusalem, Madrid, Paris, Leiden, and Oxford, were showering him with all sorts of professorships, research posts, and honorary positions—anything he wanted, in fact, just so long as he'd dignify their campuses with his presence. He had already turned down one offer from Princeton, back in 1927, but now conditions had changed. It might be time to go to America.

"Will you be in Germany this summer?" Einstein asked . . .

Next month Flexner found himself in Caputh, Germany, striding along through a cold drizzle to Einstein's country home. He arrived at three in the afternoon and stayed until eleven at night. This time he got his answer: *Ich bin Feuer und Flamme dafür,* Einstein said ("I am fire and flame for it"). It was June 4, 1932, and Albert Einstein was the Institute's first faculty member. Suddenly Flexner's dreams for the Institute took on a whole new meaning and dimension. It was as if God himself were going to take up residence at Flexner's new place in Princeton.

Of course there were still a few problems to be worked out: Einstein's salary, plus the matter of his associate, Walther Mayer. Einstein wanted $3,000 a year. "Could I live on less?" he asked Flexner. "You couldn't live on that," Flexner said.

Flexner put Einstein's salary at $10,000, which was acceptable to the physicist, but the Walther Mayer appointment was another thing. Einstein normally avoided collaborating with others—"I am a horse for a single harness," he used to say—but he had in fact written a few papers together with Mayer, an Austrian mathematician, and the two of them had visions of coming up with a unified field theory. In addition, Einstein found it useful to have an assistant to whom he could assign his more routine calculational work while he himself did the more abstract and creative theorizing. For all these reasons he regarded Mayer as indispensable.

Flexner agreed to bring Mayer to the Institute, but not to give him a separate appointment in his own name: he would simply be Einstein's assistant. Mayer, after all, wasn't really Institute material. He had written a book on noneuclidean geometry, but that was his only claim to fame, and it would certainly do nothing for the Institute's prestige to make him a professor, no matter how useful he might be to Einstein. Einstein was insistent, however, and in the spring of 1933 he wrote Flexner saying that the whole deal was off unless the Institute hired Mayer. "I would deplore it very much indeed," Einstein said, "if I were deprived of his valuable collaboration; and his absence from the Institute might even create some difficulties for my own work." So Flexner gave a separate appointment to Walther Mayer.

On October 17, 1933, Albert Einstein, his wife Elsa, his secretary Helen Dukas, and his collaborator/assistant Walther Mayer sailed into New York harbor aboard the liner *Westmoreland*. "It's as important an event as would be the transfer of the Vatican from Rome to the New World," said Einstein's friend Paul Langevin. "The pope of physics has moved and the United States will now become the center of the natural sciences."

The Einsteins disembarked at Quarantine Island. There they were met by Institute trustees Edgar Bamberger and Herbert Maass, who handed the professor a letter of welcome from Flexner. The director had arranged for Einstein's party to bypass the reception that New York City Mayor John P. O'Brien had staged for the occasion, complete with cheerleaders, parade, and speeches. O'Brien was then involved in an election campaign against Fiorello LaGuardia, and, while visions of capturing the Jewish vote withered before the mayor's eyes, Einstein was taken off in a small launch to a point on the New Jersey shore, then driven south past the scenic garbage dumps and oil refineries of northern New Jersey to his new home in Princeton.

Einstein spent the first few days at the Peacock Inn, an old rooming house a few blocks from the university. Later he moved to 2 Library Place, across from the graduate school. Finally, he purchased a white clapboard house at 112 Mercer Street, where he would live for the rest of his life. The house, which goes back at least to the early 1800s, was on a busy street. It was not the best spot for quiet contemplation, but Einstein took a second-floor room at the rear, overlooking some tall pines and oak trees, and made it his private study. Finally, he could get on with his work.

By the time he was settling down in Princeton, the theory of relativity was thirty years old, almost ancient history. Einstein had published his special theory in 1905, the general theory in 1915, and then he was done with it. But all this time he had a problem knocking around in his head that was bothering him far more than relativity ever did. It was the quantum problem. From the turn of the century until the day he died in 1955, quantum problems drove Einstein crazy. "I have thought a hundred times as much about the quantum problems as I have about general relativity theory," he said, but it was all to no avail. At the end of his life, Einstein was as baffled by them as he'd been at the very beginning, if not even more so. "All these fifty years of conscious brooding have brought me no nearer to the answer to the question 'What are light quanta?'," Einstein said. "Nowadays, every Tom, Dick, and Harry thinks he knows it, but he is mistaken."

Quantum theory is one of the most successful constructions in the history of physics. Its predictions have been verified again and again—not that this ever made an impression on Albert Einstein. "The more success the quantum theory has," he said in 1912, "the sillier it looks."

In his years at the Institute for Advanced Study, quantum theory was Einstein's private obsession. While his colleagues were turning to quantum mechanics as if it were manna from the Platonic Heaven, Einstein just shook his head in disbelief. All this stuff about the observer influencing reality, events happening randomly, causelessly—it just made no sense. It was plain unreasonable, and Einstein took every chance for saying so, trotting out his famous God-wouldn't-have-done-it epigrams: "God does not play dice with the world," "God may be subtle, but he is not malicious," and so on.

In 1935 Einstein teamed up with two of his Institute colleagues, Boris Podolsky and Nathan Rosen, to write a four-page paper that aimed to disprove quantum theory—or at least to throw a monkey wrench into the proceedings. The Einstein-Podolsky-Rosen paper—whose central argument became known to physicists as "the EPR paradox"—sent shock waves throughout the physics community. The argument was puzzling. Physicists weren't quite sure how to respond to it, except to say that it was wrong. Einstein got letters telling him that the EPR paradox was not really a paradox, that the whole thing rested on a simple misunderstanding, that it was all a mistake. Einstein was amused, though, that no two letters seemed to agree on just what the mistake was.

But the EPR episode was made all the more extraordinary by the fact that it was Einstein himself who had, in 1905, first advanced the revolutionary idea that light consisted of quanta and not, as everyone had then suppposed, of waves. So it was quite ironic that the very man who had in his youth revolutionized physics with his theory of relativity, and who had himself proposed the light quantum idea, it was ironic that he should now be in active revolt against his own intellectual offspring. He had been brought to the Institute for Advanced Study to prove what a forward-looking place it was, but virtually his first significant act there as a physicist was an attempt to overturn the theory that seemed to be the wave of the future. It was as if he were taking physics back to the Dark Ages, and other physicists were a bit distressed. J. Robert Oppenheimer visited the Institute in 1935, the same year that Einstein, Podolsky, and Rosen were coming out with their paradox. "Einstein is completely cuckoo," Oppie said at the time.

Einstein's involvement with the quantum goes back to the turn of the century, to 1900, when Max Planck, a physicist at the University of Berlin, discovered that levels of electromagnetic radiation varied in discrete, steplike stages rather than continuously. The heat of a glowing coal, for example, or the light of the sun, could radiate at one specific energy level, or at some other specific energy level, but not at any level in between the two. It was as if there were some *dead space* in there between the two levels, as if nature were somehow digital rather than analog. But why?

The whole thing was a mystery. Energy came in the form of waves—or so everyone thought at the time—and waves are smooth and continuous by definition, and so they ought to come in any arbitrary amplitude or frequency. But the empirical fact of the matter was clearly that they did not: rather, energy came in discrete permissible units—"quanta of action," as Planck called them. Planck showed that these quanta always came in whole multiples of a certain value, 6.55×10^{-27} ergs per second, a value now known as Planck's constant, and symbolized as h. Planck summed up his discovery in the equation,

$$E = h\nu,$$

which states that energy (E), is equal to Planck's constant (h) multiplied by the frequency (ν) of the radiated energy. The strange thing was that although the equation seemed to fit the experimental facts perfectly, Planck nonetheless couldn't quite bring himself to accept the conclusion that energy somehow came in discrete units. *Matter* did, of course, but not energy.

Five years later, Albert Einstein showed that energy was not being broadcast in the form of waves after all, that it was indeed coming out in the form of *particles*. On the gross level, Einstein admitted, light may appear to be continuous and wavelike; nevertheless, it actually came in the form of self-contained packages or parcels. "In accordance with the assumption to be considered here," Einstein wrote in his 1905 paper on light quanta, "the energy of a light ray spreading out from a point source is not continuously distributed over an increasing space but consists of a finite number of energy quanta which are localized at points in space, which move without dividing, and which can only be produced and absorbed as complete units."

And so the theory of light quanta was born. At the time, though, nobody could quite believe it. Einstein won out in the end, of course, although it was an uphill battle. The University of Chicago experimentalist Robert Millikan, for example, took it upon himself to *disprove* the "bold, not to say the reckless hypothesis of an electromagnetic light corpuscle." So he devoted himself to testing the theory. "I spent ten years of my life testing that 1905 [prediction] of Einstein's," he said later. "And, contrary to all my expectations, I was compelled to assert in 1915 its unambiguous verification in spite of its unreasonableness."

By 1920, physicists were finally willing to accept the notion that Einstein had proposed fifteen years earlier, that light was propagated through space not as waves, but as discrete particles.

When, in the fall of 1933, the Institute for Advanced Study at long last opened its doors to the world, it did so not in Newark, but in Princeton, specifically in Fine Hall, the university's mathematics building. Ever afterward, people have had the impression that the Institute is somehow affiliated with the university, as in the phrase, "Princeton University's Institute for Advanced Study." But this isn't correct, and never has been, for the Institute used the university's offices only as temporary quarters while it was waiting to purchase land for a campus of its own, which it finally did in the late 1930s. Meanwhile, it was located in Fine Hall, as a guest of the university, to which the Institute later made a gift of some $500,000 in return for the favor. The two institutions, however, are now and always have been financially, administratively, and organically separate, two distinct entities.

When it opened for business, the Institute had, in addition to Einstein, three other professors on its roster: James Alexander, John von Neumann, and Oswald Veblen. Together, this crew of four had enough academic brilliance to redeem Abraham Flexner's earlier claims that he had "deliberately hitched the Institute to a star," had "sketched an educational Utopia," and had founded "a Paradise for scholars." Almost from the beginning, however, there was trouble in Paradise.

For one thing, there was the embarrassing fact that three of the four had been hired away from the Princeton University mathematics department, this despite Flexner's assurances that he "would not for the world do anything to mar the great work in mathematics that is going on at Princeton." In part to atone for their sins, Flexner and Louis Bamberger promised university dean Luther Eisenhart that this would never happen again. The Institute has technically honored this informal "no-raiding agreement," although one might quibble about how well it lives up to the spirit of the thing. Years later, when the Institute had designs on Princeton mathematician John Milnor, the whole issue was neatly side-stepped.

Milnor was a mathematician at the university. He'd been a student there, then a professor, and was on the verge of becoming a permanent fixture around the place, when the Institute decided it would like to add him to its own roster of stars. Milnor, after all, was a winner of the Fields medal, the highest honor in the discipline, and the Institute collects Fields medalists like butterflies. But of course there was this no-raiding business to contend with. Well, what if Milnor were to go somewhere else for a year or two, and *then* come to the Institute? He would come there from some third institution and so technically it would be as if the Institute weren't really taking him from Princeton at all. All the parties involved could be viewed as having lived up to their obligations. And so John Milnor taught at UCLA for a year and at M.I.T. for a couple of years and then, in the fall of 1970, he was made a professor of mathematics at the Institute for Advanced Study.

Another problem way back then was the matter of salaries. Not that they were too low, just the reverse. They were too high . . . at least in the eyes of those not getting them, to wit, the unchosen and left-behind at the university. It had always been one of Abraham Flexner's primary aims to remunerate his staff so lavishly that a professor at the Institute would never have to "eke out his inadequate income by writing unnecessary textbooks or engaging in other forms of hack work." The whole point of the place, after all, was to take care of its members' usual terrestrial needs to such a degree that the only conceivable remaining activity was thinking. But in handing out salaries Flexner was a wanton overachiever, offering Oswald Veblen $15,000 per annum (a full $5,000 more than he gave Einstein, at least to start off with), plus a retirement pension of $8,000, *plus* a lifetime pension of $5,000 for Veblen's wife. In the 1930s, these figures were astonishing. Veblen's retirement pay alone was equal to, and in some cases greater than, the normal full-time salaries of some of Princeton's best professors. This rankled. As the Institute's historian Beatrice Stern commented years later, "The shining example is hard to live with, especially when it is the same old colleague with a new hat."

But there were also problems within the Institute itself. Flexner had all along described the place as an institution that would bestow "the degree of Doctor of Philosophy and other professional degrees of equal standing"—this was in fact stated in the Institute's certificate of incorporation.

Suddenly, and without explanation, Flexner announced that "only those students would be admitted who have already obtained the Ph.D. degree or whose training is equivalent to that represented by the Ph.D. degree and who are in addition sufficiently advanced to carry on and to cooperate in independent research." In other words, the Institute would not offer the Ph.D. degree after all.

This change came as a distinct surprise to Oswald Veblen, who was head of the Institute's School of Mathematics. In retrospect, though, perhaps it shouldn't have. In the summer of 1932, Flexner had taken Veblen on a field trip to New York, to give him a tour of the Rockefeller Institute for Medical Research, where his brother, Simon Flexner, was director. Flexner explained to Veblen that the Rockefeller Institute did not award degrees, that its entire *raison d'être* was research, and that this was also to be the case at the Institute for Advanced Study. And a year later, shortly before the Institute opened, Flexner had written a letter to Veblen, saying, "I don't want to begin giving the Ph.D. degree, for I don't want to involve the staff in theses, examination, and all the other paraphernalia. There are plenty of places where a man can get a degree. Our work must be beyond that stage."

Despite all this, Veblen admitted two students to the Institute as candidates for the Ph.D. degree, one of them having only a bachelor of science degree to his credit. This action enraged Flexner, who went to the board of trustees and had them affirm his new no-Ph.D. policy in the face of Veblen's disobedience. Flexner explained to the trustees, although evidently not to his own faculty, that although for legal reasons the Institute had applied to the New Jersey Board of Education for the authority to grant Ph.D.s, it had never been his intention to award any. To this day the Institute has never bestowed degrees, although there are a few faculty members who think it ought to.

These problems were part of the young Institute's growing pains. Still, the world's best young scientists flocked there in droves. Kurt Gödel and Alonzo Church, both logicians, came on as "workers," as they were called, as did mathematicians Deane Montgomery, Boris Podolsky, and Nathan Rosen. Einstein and Podolsky, together with Richard Tolman, had written a paper in 1931 when all three of them were faculty members at the California Institute of Technology, in Pasadena. The short, two-page piece, published in *Physical Review*, exposed an "apparent paradox" in quantum mechanics. Now at the Institute, and at the age of fifty-six, Albert Einstein got ready to make his best effort, his last great do-or-die offensive upon the theory that stirred him more deeply than anything else in physics.

In making his assault, Einstein was not by any means abandoning his own earlier theory of light quanta. He was only rejecting the later additions to quantum theory, a collection of doctrines which went under the name of "the Copenhagen interpretation." According to this new view, which had been developed by Niels Bohr and Werner Heisenberg, the observer has to be brought into the quantum picture in a fundamental way. It's meaningless, the two men asserted, to talk of the fine structure of matter without specifying the instruments and means by which observations of quantum phenomena are to be made. Bohr therefore tried quite deliberately to blur the line between the measuring instrument and the object measured: "The finite magnitude of the quantum of action," he said, "prevents altogether a sharp distinction being made between a phenomenon and the agency by which it is observed."

The reason for this was that the act of observation changes the object. As physicist Pascual Jordan put it, "Observations not only disturb what has to be measured, they produce it. . . . We compel [the electron] to assume a definite position. . . . We ourselves produce the results of measurement." Or, as John Wheeler later expressed it, "No phenomenon is a *real* phenomenon until it is an *observed* phenomenon."

But Albert Einstein didn't even like to *hear* this stuff. "When a mouse observes," he used to ask, "does that change the state of the universe?" For him, things out there in the world had whatever properties they had, and they had them whether or not you were looking. This held true on the large scale, and he wanted it to be true on the small scale as well, on the scale of quanta. For Einstein, no technical scientific doctrine could override the more fundamental philosophic notion of "objective reality," the principle that things possess all their properties independent of and prior to the act of observation. For Einstein, the act of observation creates no properties.

Here, at least, Einstein was no relativist. "We often discussed his notions on objective reality," said Einstein's biographer, Abraham Pais, who knew him at the Institute. "I recall that during one walk Einstein suddenly stopped, turned to me and asked whether I really believed that the moon exists only when I look at it."

Einstein thought he had found a serious paradox in quantum theory and he wanted to develop it further, so he discussed his ideas with his old colleague, Boris Podolsky, and with the twenty-six-year-old Nathan Rosen, who had recently gotten a Ph.D. in physics from M.I.T. Collectively, they would argue that on quantum theory's own assumptions, there must be something in nature *beyond* what appears in experimental observations. There must be some underlying reality that is persistent and stable. Quantum theory did not acknowledge any such an underlying reality—in fact it expressly denied it—which meant that quantum theory was *incomplete* as a full account of nature. Thus they entitled their article, "Can Quantum-Mechanical Description of Physical Reality Be Considered Complete?," and they answered with a resounding No.

Fifty years later, their argument is still controversial. The three collaborators took as their starting point the Heisenberg "uncertainty principle." Heisenberg had discovered that quantum attributes come in pairs, such as position and momentum, or energy and elapsed time, and that these pairs—"conjugate variables," they're called—are so related that you cannot know them both with perfect accuracy in a single experiment. He stated this as a mathematical relationship, according to which, if Δx represents the uncertainty of one attribute (say, position), and Δy the uncertainty of another attribute (say, momentum), then the product of these uncertainties is greater than or equal to Planck's constant, h:

$$(\Delta x) \times (\Delta y) \geq h.$$

Despite its name, the uncertainty principle does not maintain that "everything is uncertain" at the quantum level. In fact it asserts just the opposite, that any *one* attribute of a quantum particle can be known with complete and utter exactness. The price you pay for this exactness, however, is that all knowledge of the particle's *other* paired property will be lost. For example, you may know an electron's position with certainty, but then, according to the uncertainty principle, you can know nothing whatever about its momentum. As Werner Heisenberg once put it, it's like "the man and the woman in the weather house. If one comes out the other goes in." The thing that prevents us from having exact knowledge of both values simultaneously is not any metaphysical bashfulness on the part of quanta, but rather is the basic quantum-mechanical fact that the act of measurement disturbs the object measured. A particle once observed is a particle lost from further view.

Einstein, Podolsky, and Rosen argued, however, that quanta can be understood to have properties that are as definite and objective as any in classical physics. Suppose, they said, that you had two particles, A and B, and that you knew their total momentum and relative position. Now this is something that orthodox quantum theory allows for with no problem: it allows that the sum of the momentums of two particles, as well as their relative positions, can be known with certainty. So let us assume, they argued, that the two particles interact, and that they then separate and fly apart, going off to great distances, until they're perhaps light years away from each other. From the law of conservation of momentum we know that the combined momentum of the two particles is the same both before and after their interaction. But Einstein and his collaborators now noticed that if you measured the momentum of either one of the two particles, then not only would you know that particle's momentum, you'd know the other particle's momentum as well. And you'd know it *without having disturbed the other particle in any way*. This was significant, for it meant that a quantity was there whether or not you measured it, something that quantum mechanics steadfastly denied was possible.

Say, for example, that the total, combined momentum of particles A and B is 10 (in some units), and that after the two particles separate A's momentum is measured and found to be 6. Subtracting 6 from 10 we would know that particle B must have a momentum of 4, and we would know this without disturbing B at all. But if you can learn the momentum of a particle without disturbing it, then the particle must have that momentum *whether or not* you take any steps to measure it. In other words, its momentum is something that must exist "objectively," apart from observation.

But the corresponding argument can be made concerning the relative positions of the two particles. Measuring the position of one of them will allow you to deduce the position of the other one without disturbing it. But in that case the two particles must have possessed their respective positions prior to and independent of any act of measurement, which means that position is as objective an attribute as momentum. So as far as Einstein and his associates were concerned, both these quantum properties were objective realities existing apart from measurement. But since quantum mechanics didn't allow for objectively existing properties, Einstein, Podolsky, and Rosen declared that quantum mechanics is fundamentally incomplete as a theory of nature.

To Niels Bohr, all this was distasteful in the extreme. He didn't much like the notion of "objective reality," and in fact had spent a good part of his career as a physicist undermining that very concept. Bohr wanted to see it replaced by an alternative notion called "complementarity," according to which reality and knowledge are inseparably intertwined and, in which, as he described it, "no sharp distinction can be made between the behavior of the objects themselves and their interaction with the measuring instruments." For Einstein and his friends to claim that you could separate reality and knowledge from each other like back in the old days—back in the halcyon days of classical physics—this was a throwback that had to be nipped in the bud. The question was, how?

"This onslaught came down upon us as a bolt from the blue," says Leon Rosenfeld, who was in Copenhagen with Bohr when news of the EPR paradox arrived. "Its effect on Bohr was remarkable," he said. "As soon as Bohr had heard my report of Einstein's argument, everything else was abandoned: we had to clear up such a misunderstanding at once." The EPR argument even made the newspapers. On May 4, 1935, the *New York Times* carried a story about the dispute: "Einstein Attacks Quantum Theory," the headline read.

Bohr, greatly excited, immediately began to dictate a reply to Einstein and company. He found, however, that this was no easy matter. He'd start off on one track, then change his mind, backtrack, and start again. He couldn't put his finger on exactly what the problem was. "What *can* they mean? Do *you* understand it?" he asked Rosenfeld.

After some six weeks of work, Bohr had an answer. Calling for "a final renunciation of the classical ideal of causality," for a "radical revision of our attitude as regards physical reality," and for a "fundamental modification of all ideas regarding the absolute character of physical phenomena," Bohr insisted that a measurement of one particle *does* in fact affect the other one in some unspecified way, and that a correct understanding of quantum phenomena will properly include the influence of measurement upon both particles under consideration.

Today the correctness of the EPR argument remains an undecided question for physics. In a review of the literature that has piled up during the interval, Cornell physicist David Mermin wrote in 1985, tongue not entirely in cheek: "Contemporary physicists come in two varieties. Type 1 physicists are bothered by EPR . . . Type 2 (the majority) are not, but one has to distinguish two subvarieties. Type 2a physicists explain why they are not bothered. Their explanations tend either to miss the point entirely . . . or to contain physical assertions that can be shown to be false. Type 2b are not bothered and refuse to explain why. Their position is unassailable. (There are variants of type 2b who say that Bohr straightened out the whole business, but refuse to explain how.)"

In the early 1980s, experimenters Alain Aspect and collaborators at the Institute of Theoretical and Applied Optics of the University of Paris performed a test that seemed to show that if you take the EPR argument seriously then there could be instantaneous signaling across great distances, a circumstance that led one researcher to suggest to the United States Department of Defense that EPR correlations could be used as a method of faster-than-light communication among the nation's submarines. Whatever the merits of this proposal, it would be amusing to Einstein that, more than fifty years after he came up with his most famous challenge to quantum mechanics, the EPR argument still unsettles some physicists.

In the interim, of course, many physicists wrung their hands over Einstein's continued opposition to the theory he had helped create. "He was a pioneer in the struggle for conquering the wilderness of quantum phenomena," said Max Born in 1949. "Yet later, when out of his own work a synthesis of statistical and quantum principles emerged which seemed acceptable to almost all physicists, he kept himself aloof and skeptical. Many of us regard this as a tragedy—for him, as he gropes his way in loneliness, and for us who miss our leader and standard-bearer."

But the results of Einstein's lone groping are still with us today. Richard Feynman said of the EPR paradox in 1982, "I cannot define the real problem, therefore I suspect there's no real problem, but I'm not sure there's no real problem."

Objective reality, in other words, may still have a chance.

Einstein's intellectual battles with quantum theory were paralleled by others he was waging against Abraham Flexner. From the very beginning, Flexner wanted the Institute to be remote and withdrawn, as cut off as possible from the rest of the world. "It should be a haven," he said, "where scholars and scientists could regard the world and its phenomena as their laboratory, without being carried off in the maelstrom of the immediate." Not everyone who watched the Institute take shape, however, was equally convinced that this withdrawal from the external world was a good thing. Dr. George E. Vincent, of the University of Chicago, told Flexner that, at Chicago at any rate, people at the upper reaches were not cut off from reality, that on the contrary they placed themselves "in the maelstrom" and emerged from it all the healthier. Arnold Toynbee, likewise, told Flexner that the utter detachment he contemplated might in the end lead to intellectual sterility. People must be of their times, Toynbee said, and he advised

Flexner not "to cut their roots." (Toynbee evidently felt able to cut his own roots now and then, for he was an Institute member on five separate occasions during the late 1940s and early '50s.)

Flexner was intent on keeping his whole faculty from immersion in things, but he made some special efforts in the case of Einstein, for he did not want the man turned away for a moment from his assignment of producing further revolutionary physics. Before Einstein ever arrived at the Institute, letters, telegrams, and telephone calls meant for him started pouring in, and of course Flexner had these intercepted, and even answered, on Einstein's behalf. (The Institute even now sometimes receives letters addressed to Albert Einstein.) The trouble was, Flexner kept on with this even after Einstein had arrived on the premises.

Shortly after the Institute opened, a call came in from Marvin MacIntyre, secretary to Franklin D. Roosevelt, president of the United States. The president would like to invite Einstein and his wife to dinner at the White House. The call somehow got through to Einstein's secretary, who accepted the invitation. Flexner heard about this and called the White House back, informing them—and none too politely—that Einstein's appointments could be made only through him (Flexner), and that unfortunately the professor could not come to dinner after all. Flexner followed this up with a stern letter in which he explained that "Professor Einstein has come to Princeton for the purpose of carrying on his scientific work in seclusion," adding that "it is absolutely impossible to make any exception which would inevitably bring him into public notice."

Ultimately, of course, Einstein straightened everything out and got himself to dinner at the White House. Flexner, though, kept on as if he were Einstein's own press secretary and personal public relations manager. After a while Einstein developed a sense of being held hostage at the Institute, and so when he wrote letters to close friends he'd put down a return address of "Concentration Camp, Princeton." It finally got to the point where he complained to the Institute's board of trustees. Flexner, he charged, "has interfered several times in my private affairs, and that in a very tactless way. . . . He has written insulting letters to my wife and myself." Moreover, Flexner had tried to stop Einstein from appearing at the Royal Albert Hall in London, had intercepted important letters and telegrams, and so on and so forth. If Flexner didn't mend his evil ways, Einstein would resign from the Institute. As always when faced with the prospect of one of his star faculty members defecting, Flexner relented.

Einstein's difficulties, though, were not confined to his dealings with Abraham Flexner. He also had to find a replacement for his once-trusted assistant, Walther Mayer. It seems that Mayer, no sooner had he arrived at the Institute, put a great distance between himself and his master. Their work together consists only of a single paper published in 1934, after which Mayer returned to his own work in pure mathematics, refusing any longer to be associated with the theory of the unified field. In 1936/37, Einstein took on two new assistants, Peter Bergmann and Leopold Infeld. He wanted them to continue on for the academic year 1937/38 as well, but he immediately ran into problems getting their stipends renewed. In the School of Mathematics, money was controlled by Oswald Veblen, who unfortunately took the position that since Mayer had been brought to the Institute specifically to aid Einstein, Einstein was therefore entitled to no other assistant. Einstein appealed directly to Flexner, who saw to it that Bergmann got a stipend for the next several years. The case of Infeld, though, was left hanging.

In February of 1937, Einstein asked his colleagues to appropriate Infeld the modest sum of $600 for academic year 1937/38. In an uncharacteristic move, Einstein attended a meeting of the school's faculty to make a special appeal on behalf of Infeld, and, to his astonishment, got nowhere. "I tried my best," he told Infeld at the time. "I told them how good you are, and that we are doing important scientific work together. But they argued that they don't have enough money . . . I don't know how far their arguments are true. I used very strong words which I have never used before. I told them that in my opinion they were doing an unjust thing. . . . No one helped me."

Einstein thereupon offered to pay Infeld the $600 out of his own pocket, but Infeld declined. In desperation, he proposed writing a popular account of the evolution of physics, to be published and marketed with Einstein listed as co-author. Einstein agreed, and Infeld wrote the book during the summer of 1937. When *The Evolution of Physics* came out in 1938, it earned its authors a lot more than $600, but bitter memories of the Institute's stinginess lingered on long afterward.

Finally there was the Erwin Schrödinger affair. Schrödinger, who had just won the Nobel prize for physics in 1933 (jointly with P.A.M. Dirac), had worked together with Einstein at the University of Berlin in the 1920s. Schrödinger came to Princeton in the spring of 1934, to take up a visiting professorship at the university. He and Einstein, both of whom seemed to be on the same wavelength in physics, worked together again and got along quite happily. The university offered Schrödinger a permanent position, which he declined on the grounds that he expected that the Institute would offer him an appointment in the immediate future, and that he would prefer to work there instead. The Institute, however, had no plans to appoint Schrödinger, and apparently never had. Despite Einstein's once again appealing to Flexner, the Institute made no offer to Schrödinger, who ultimately returned to Europe.

To explain this somewhat shabby treatment of Einstein, some have suggested that Veblen, who was a pure mathematician, bore some animus toward Einstein, who was largely a physicist. Another story has it that the physicists already on the faculty held it against Einstein that he was engaged in dismissing quantum mechanics at every turn. A third version is that everyone was disappointed that the older Einstein got, the less contribution he made to physics, so that they in turn ceased to make special efforts on his behalf. After all, Einstein was serving his major function just by *being there* at the Institute. He didn't really have to do anything: his role was now to be that of a figurehead, a living icon, a patron saint. "He is a landmark, but not a beacon," Oppenheimer would say later.

Einstein learned what his role was soon enough. The Institute has an annual ball every spring, and on the morning of the dance in the last year of Flexner's directorship, Einstein mentioned to his assistant Valentine Bargmann that he would see him again at the dance. Bargmann was surprised that Einstein would be going. "Of course I'm going," Einstein said. "I'm taking this very seriously because that's what Mr. Flexner bought me for."

In 1936 the Institute purchased Olden Farm and some adjoining lots, altogether amounting to a 200-acre tract of land about a half mile south of the Princeton University campus. Three years later, after getting another large gift from Louis Bamberger and Caroline Fuld, the Institute opened for business in Fuld Hall, a brand new four-story red-brick building that had something for everyone: offices for the director, faculty, and temporary members, a mathematics library, and a common room for afternoon tea.

Everyone at the Institute welcomed the departure from the Princeton University campus. Some, Einstein included, felt that the university's president, Harold Dodds, had too much influence over Flexner. At a faculty meeting in the late 1930s, Professor James Alexander reported that "a young mathematician was denied admission to the Institute, in a discreet manner, because being colored Princeton University would have objected to him." What with renewed charges of anti-Semitism at Princeton, all concerned thought it was good for the Institute finally to be in control of its own destiny.

Abraham Flexner, however, never made it to his new offices, for, long before the new building opened, the faculty was already plotting to unseat the director and replace him with Institute trustee Frank Aydelotte. At issue was what many professors regarded as Flexner's mania for keeping them out of the Institute's administrative affairs. Originally, Flexner wanted to give his faculty members some say in the running of the place, but he finally decided that self-governance would only lead to endless bickering. But Flexner himself provoked all the bickering he would ever want to see when in the late 1930s he moved to hire two new men in economics.

This led to Faculty Mutiny Number One.

At its inception in 1933, the Institute consisted only of a School of Mathematics. He began with mathematics, Flexner explained, because "its devotees are singularly unconcerned with use, most of all with immediate use, and this state of mind and spirit, it seemed to me, ought to dominate the new institute." Despite this Platonic-heavenly emphasis on pure theory, however, the Institute added on a School of Economics and Politics two years later. Partly this reflected Flexner's own prejudices, for he thought that economists could improve the state of the world. But partly it was in response to advice that Flexner had gotten earlier, from people like the historian Charles Beard who, in 1931, urged him to *begin* with economics. "Chuck mathematics and take economics," Beard told him. "Then you begin with the hardest subject. It is as mathematical and statistical as anyone wants to make it, but it is more. It is a far more 'severe' discipline than mathematics, because it deals with the inexact." The School of Economics and Politics, at any rate, opened with three professors, Edward Mead Earle, David Mitrany, and Winfield Riefler.

Later, when Flexner wanted to add some new men to the department, he tried to do so without consulting two of the three men already on the staff. This gave the impression that he was railroading people in for his own purposes, an impression that was not dispelled when the rest of the faculty saw who it was Flexner was hiring, to wit, Walter Stewart and Robert Warren. These gentlemen were not Institute material by any stretch of the imagination. They just weren't scholars. Not only did they both lack Ph.D. degrees, one of them had only a bachelor's degree from . . . the University of Missouri! It was true that both of them had done some work for the government, but did *this* qualify them for the Institute? Somehow, apparently, Flexner had gotten it into his head that it did, that their experience out there in the world at large would give them a unique vantage point up here in the ivory tower. They might even be able to come up with a scheme for curing for the Depression. Flexner even began to speak of a "clinical economics" program at the Institute.

Plainly, Flexner was losing his marbles. The Institute was supposed to be the last bastion of scientific theory, and here was Flexner, trying to bring in "men of affairs." What's more, the Institute is supposed to hire only the elite, people whose degrees are marked all over with honors, distinctions, and *summa cum laude,* and here was Flexner, bringing in people who could hardly be hired at a third-rate Bible college in Arkansas. Worse still, Veblen—who was faculty representative to the board of trustees—now went around telling other faculty members that in a trustees meeting Flexner had claimed that the Stewart and Warren appointments would be temporary but later recorded them in the minutes as permanent.

This was serious. Clearly, Abraham Flexner had to go.

Albert Einstein had by this time lost all patience with the director's pettiness, meddling, and bad judgment. At a meeting in the Nassau Tavern, Einstein presided as host and chairman while a plan for Flexner's removal was hatched. Later, in the spring of 1939, Einstein, together with mathematician Marston Morse and archaeologist Hetty Goldman, wrote a letter to Flexner asking that the Institute's faculty be consulted on any new appointments for professor or director. Flexner met with Einstein and Morse, but didn't say yes or no to their request. The fact of the matter was that by this time a new director, Frank Aydelotte, had already been chosen. Flexner, who was aware of this, nevertheless did not impart the information to any of his critics, who continued to plot his downfall.

By the time he left the directorship, Abraham Flexner had become an embittered and resentful man. He gave his successor, Frank Aydelotte, a few parting words of advice. "Don't, for your own sake and that of the Institute underestimate the fact that you are dealing with intriguers," he said. "I freely confess that I was a baby in their hands. I took them at their word; I supposed that when they said they wanted opportunities for scholarship and wanted to be free of routine they meant it. They did not mean a word of it—that is, a few or several of them. They wanted opportunities for scholarship, with high salaries, but they also wanted managerial and executive powers. They saw they could not get them through me directly;

hence Veblen and a few others intrigued to get them indirectly. . . . Veblen wants power. Maass wants importance. You will have to make them both realize from the jump that you are the master."

And with this injunction from on high, Abraham Flexner retired as director of Princeton's Paradise for Scholars. Faculty Mutiny Number One had succeeded. There would be others.

In the fall of 1939, as Frank Aydelotte was coming in as director, Albert Einstein settled into his new office in Fuld Hall, room number 115. It was a large, airy chamber at the back of the building, fitted out with blackboard and bookshelves, an oblong meeting table at one end, and a bay window at the other end that let in a lot of light.

Light. It was Einstein's specialty, almost his own personal domain, the topic that had fascinated him since, at the age of sixteen, he asked himself what the world would look like if you traveled on a light wave, and then, during "the miracle year," 1905, formulated the special theory of relativity which defined the speed of light as one of the absolutes, or invariants, of the physical universe. In that same year he proposed that quanta of light explained the photoelectric effect, and in 1911 he predicted that light rays would be bent by gravity. In 1917 he advanced the idea of a light photon, a fundamental, massless, pointlike parcel of energy and momentum. Einstein, in short, had done more than any other single physicist to understand the nature of light. Now, toward the end of his life, the physicist worked on what he considered to be his final and ultimate task, that of uniting the phenomena of light and gravity into a single overall theory.

A unified field theory, if it existed, would be one of the great intellectual achievements of mankind, comprehending all the disparate elements of the universe into a single overarching law. Some important unifications had been made before Einstein. Newton, for one, combined terrestrial and celestial gravity into a theory that encompassed them both. (Prior to Newton it was not known that objects on or near the earth's surface—like artillery projectiles, for example—were governed by the same force, gravity, that held the planets in orbits and stars in galaxies.) The next great reduction was achieved by Maxwell in his equations for electromagnetism, which brought together electricity, light, and magnetism under the same set of differential equations.

When Einstein set to work to unite all known forces into a single unified field theory, he had in front of him Newton's laws, Maxwell's equations, his own theory of relativity, as well as the laws of quantum mechanics. In seeking a unified theory of all nature, Einstein wanted to show how all the universe's diverse forces and particles could be comprehended under the grasp of one consistent set of principles.

This is a tall order. For one thing, Einstein had no proof that such a unification was possible, or that it even reflected the facts. It might just turn out, for example, that the world's phenomena are *not* in fact subsumable under any one law at all, that, to the contrary, they are at base different and essentially unrelated phenomena. Another obstacle was that Einstein had no assurance that, even if such an all-embracing law existed, it could be comprehended by the human mind. But he'd done the impossible before, and maybe he could do it again, simply by thinking.

Einstein completed his first paper on the unified field theory in January 1922. Working on the problem, he had made many false starts, backtracking, starting all over again, taking up again and again approaches he'd long since discarded. He did this until the very last day of his life. At the Institute Einstein pursued the unified field theory with Walther Mayer, Valentine Bargmann, Peter Bergmann, Ernst Straus, Bruria Kaufman, and a host of other collaborators and assistants. His usual working day would begin at 9:30 or 10 in the morning, when he would leave his 112 Mercer Street home and walk the mile or so to the Institute. Oftentimes he would be accompanied by Kurt Gödel, who lived even farther away, and who would stop by to collect Einstein on his way to work. Arriving at the Institute, Einstein would meet in his office together with his assistants to discuss any ideas that seemed at the moment to offer some promise. They would work at this for a couple of hours, then Einstein would return to his home for lunch, after which he'd resume work on his own. If during the afternoon he happened to discover anything interesting, he would telephone one of his assistants and give him the good news. These were always false leads, of course, because Einstein never found his theory of the unified field. Of his last field equations, written a few months before his death, Einstein said, "In my opinion, the theory presented here is the logically simplest relativistic field theory which is at all possible. But this does not mean that nature might not obey a more complex field theory."

The Institute for Advanced Study was in the end a place of both success and failure for Einstein. He wanted to get beyond the disorder of quantum mechanics and find a more stable reality beneath the observed phenomena, and here at least he was partially successful: the EPR paradox has a life of its own today. He wanted to unify gravity and electromagnetism, and he certainly failed at that. And he wanted to install a world government upon earth that would ensure world peace, and he failed at that, too. Much of Einstein's later years were devoted to producing paralyzingly boring and monumentally naive political tracts, but in any case Einstein never regarded politics as all that important. Once when he was walking the Institute grounds with his assistant, Ernst Straus, Einstein commented, "Yes, we have to divide up our time like that, between our politics and our equations. But to me our equations are far more important, for politics are only a matter of present concern. A mathematical equation stands forever."

Einstein, of course, was in the Platonic mold as much as anyone at the Institute. The real world is out there, but the life of the mind transcends it, and this was the whole point of his science. "I believe with Schopenhauer that one of the strongest motives that lead men to art and science is escape from everyday life with its painful crudity and hopeless dreariness, from the fetters of one's own ever-shifting desires. A finely tempered nature longs to escape from the personal life into the world of objective perception and thought."

After he died, photographers came into room 115 at the Institute for Advanced Study to take the famous pictures, those shots of the cluttered office as Einstein left it before going into the hospital. The great man's chair is empty, and the room has all the appearance of a museum exhibit.

It isn't, though. After Einstein moved out, Institute astronomer Bengt Strömgren moved in and used the office for ten years, during which time he inaugurated the Institute's astrophysics program. There is no evidence that Strömgren was in the least intimidated by the thought of Einstein's ghost looking over his shoulder. After Strömgren returned to Denmark, Einstein's office went to mathematician Arne Beurling, who remains there today.

Beurling loves Einstein's office, particularly late in the day, when the sun, now lowing toward the horizon, shines in through the large bay windows at the end of the room. That's when three fingerlike shafts radiate into the chamber, flooding its darkest recesses with the object of Albert Einstein's truest love, light.

Chapter 3

The Grand High Exalted Mystical Ruler

January, 1978. A frail old man in hospital clothes is sitting up in a chair in Princeton Hospital. Deep, sunken eyes stare out through heavy round glasses at the winter afternoon. The man is slight and emaciated, and weighs only about eighty pounds. He's been in and out of hospitals, sanatoria, and doctors' offices time and again during the course of his life, for nervous breakdowns, depressions, many imagined ailments, and a few real ones. Always private and reclusive, he's regarded as an exceedingly strange man by those who know him; some say that he's been mentally unstable since childhood. The patient has a bladder problem, but he has refused any treatment for his condition despite what at least two urologists tell him. But this is not why he's so thin. The fact is that Kurt Gödel, the world's greatest living logician, probably the greatest since Aristotle, and one of the most renowned figures at the Institute for Advanced Study, second only to Einstein, abjectly refuses to eat. Gödel believes that his food is poisoned, and that his doctors are trying to kill him.

March, 1982. All John Dawson wanted to do was to get copies of a few of Kurt Gödel's unpublished papers. Dawson was a mathematician at Penn State and had written to the Institute time and again—to Harry Woolf, director of the Institute, and to Atle Selberg and Deane Montgomery, of the School of Mathematics—asking for copies of one thing and another, but he always got the same answer: Gödel's papers have not been catalogued, and so we cannot permit access to the papers, and we are very sorry, and so on. But Dawson kept trying. He made it a point to pester them up there on a regular basis. After all, Gödel was to mathematics what Einstein was to physics. You couldn't just leave the man's lifework, his intellectual *corpus*, moldering down in some basement where it's going to vanish into the mists of time.

But one day, out of the blue, Dawson gets a call from Armand Borel at the Institute. Borel is one of the mathematicians of the age, rather imposing, and it's a little unnerving just talking to him. But anyway Borel is saying that, since the Institute is interested in getting Gödel's papers catalogued—"archived," as the expression was—and since you're so interested in them, well, why not come up here and look them over and see if you'd like to take on the job?

This was a bit of a shock to Dawson. Nevertheless, a week later he's in the basement of the Institute's library, waiting for a secretary to open the door to one of the gray wire-mesh archival cages, the one with all Gödel's stuff in it. Dawson is feeling some pangs of apprehension at this moment, for the fact is that he's about to see the innermost private papers of a man whose life is almost a complete blank. Despite his having been the author of one of the most famous theorems of logic and mathematics, no one seems to know the first thing about Kurt Gödel as a person. It was the direct opposite of Einstein's case, where everyone already seemed to know everything, whether they wanted to or not. It was like it was part of your primary education; you just grew up knowing that Einstein never wore socks, that there were apparently no combs in the house, that he played the violin, and so on. There was no end to the personal anecdotes about the guy. But Gödel, that was a completely different story. Who was he? Where'd he come from? What was he like? Even John Dawson, who, like Gödel, was a professional mathematician, didn't know, and he had a hard time finding out.

"There seemed to be so few definite facts about him. I remember having an awful amount of trouble—before I went to the Institute—finding out just basic information, whether he was married, did he have any kids, and so on. *Why can't I find out this stuff?*," Dawson wondered. *"It's just basic information!"*

There were the rumors, of course—that Gödel was a hypochondriac, that he used to wear layers of sweaters and rubber galoshes on the hottest days of July—but it was hard to pin any of this down, and to find it in print was close to impossible. Even Douglas Hofstadter's *Gödel, Escher, Bach,* an incredible 700-page tome that explains Gödel's theorem and its significance for art, music, and the future course of Western civilization, says absolutely nothing about the man himself. Although his discoveries provide the intellectual center for Hofstadter's book, the author identifies him in the text only as "K. Gödel."

So it's quite understandable that a bit of excitement is welling up inside John Dawson's breast as the secretary jangles the keys to the archival cage and swings the door open in front of him. The Gödel papers are right over there, up against the far wall. It's pretty dark back there, but you can see well enough. There are two tall file cabinets and then stacks and stacks of cardboard boxes . . . there must be about sixty of them. They're piled up six feet high, in row upon row stretching across the room.

John Dawson wonders to himself: *What in the world's in all those boxes?*

Whatever else you can say about Kurt Gödel, the fact is that at the One True Platonic Heaven, he was the ruler of the roost. Not that he had any special power or authority there, just the opposite. He was neglected, passed over, and for most of the time completely ignored, both by administrators and other faculty members. It was twenty long years between the time he first came to the Institute, in 1933, and the time he was made a professor, in 1953, a delay that is unmatched before or since. Gödel was the ruler, though, in the only way that really counted at the Institute for Advanced Study, which is to say, intellectually—using degree of abstraction as the criterion for ranking. The Institute, after all, is the home of theory, and the more abstract the theory the better, and by this standard Gödel was the hands-down winner. For one thing, he was trained as a mathematician, and mathematics is the most ethereal and abstract of the sciences. It's concerned with things—such as numbers, shapes, and abstract relationships—that don't exist as such in the real world. A number, for example, is not something perceivable, nothing that you can see, hear, or hold in your hand. You can see five apples, five people, or five horses but you can't see the number five. Oh, you can see the *numeral* five—"5"—but that's not the genuine item. That's only a written expression, a symbol that *stands for* the number five. The same with geometrical shapes: triangles, circles, spheres. You can't see them, either. True enough, you can draw a triangle, but that's not a *real* triangle: it's only a *picture* of a triangle. True geometrical triangles—made up of lines of perfect straightness and zero thickness—they don't exist anywhere on the face of the earth.

But that doesn't mean they don't exist at all. Numbers and geometrical objects exist at least in the mind (because that's what we're thinking about right now), but they can't exist *only* in the mind. They exist somewhere else, too. If they were nothing but mental concepts, then numbers could be changed at our pleasure. But they can't be changed one whit. Numbers are rigid and unbending, and their properties remain the same regardless of how we may try to change them. Two plus two will always equal four, no matter what. In Gödel's view, this is explained by the fact that—even though you can't find them anywhere in nature—numbers and lines and other mathematical entities *do* have an objective existence. "Classes and concepts may . . . be conceived as real objects," Gödel said, "existing independently of our definitions and constructions. It seems to me that the assumption of such objects is quite as legitimate as the assumption of physical bodies and there is quite as much reason to believe in their existence."

All of which means that Gödel was a Platonist in the most unabashed, explicit, and literal sense of the word. He thought that mathematical objects—numbers, sets, geometrical structures—are really and in fact *out there*. Gödel never said exactly *where*, he never actually *said* that they existed up there in the Platonic Heaven, but he didn't have to. Where else could they be? And anyhow actually coming out and saying something like that—that mathematical objects exist off in another dimension—this was not considered quite acceptable, even among the other Platonists at the Institute for Advanced Study. Some things, after all, are better left unspoken. On top of it all, Kurt Gödel was already quite isolated enough, and already suspected of some kind of lurking craziness. "I am in rather lonely work," he used to say. "I am concerned with the objective existence of mathematical objects." So he didn't need to come right out and say it, but more than anyone else at the Institute, Gödel was a true believer in the World of the Forms.

When, just once, Gödel came down from the high lonesomeness of the Platonic Heaven to consider the nature of the physical world, it was to deny the reality of time and change. Now to think about time is not to get very far into the world of *things* to begin with: time is already a shadowy, insubstantial will-o'-the-wisp, as invisible as numbers or abstract shapes. Einstein, in fact, had already gotten rid of the notion of simultaneity, which made time seem even more arcane and unreal than it had seemed to begin with. But Gödel went the rest of the way: he suggested that time did not really exist in any objective sense. It's not really out there in the world at all; it's just our special mode, our own particular human way of perceiving the world.

So numbers are real but time isn't. With views like these is there any doubt that Gödel would rise to the very top—the truly loftiest estate, the utter pinnacle—of the invisible and unspoken intellectual hierarchy in place at the Institute for Advanced Study? How natural then, with Gödel ruling the roost from his position of High Lonesomeness, that he should acquire a reputation for secretiveness, an aura of mystery. How logical that he should come to be regarded as utterly profound and inexpressibly deep, as if his every word, glance, or gesture held some unsayable meaning, almost as if it were a message from the beyond.

Rudolf Rucker, a mathematician, visited Gödel at the Institute in 1972. "Listening to him," Rucker says, "I would be filled with the feeling of perfect understanding. He, for his part, was able to follow any of my chains of reasoning to its end almost as soon as I had begun it. What with his strangely informative laughter and his practically instantaneous grasp of what I was saying, a conversation with Gödel felt very much like direct telepathic communication."

And there you have it. "Perfect understanding" . . . "informative laughter" . . . "telepathic communication." Of course! Why not? We're not talking about a man, after all, a mere mortal. We're talking about the Emperor of the Forms, the Grand High Exalted Mystical Ruler.

Down in the library basement, John Dawson was walking his fingers along the last lingering traces of Kurt Gödel. "It was just like a grab bag. It was an amazing experience. I would open up an envelope and wonder *What's going to be in here*? It was really very exciting."

He'd find books, journals, manuscripts, and countless personal records of every description, including family photographs, clothing bills, apartment rental contracts, housecleaning bills, everything, anything. "There was an envelope of receipts from his wedding, right down to a counter check for a glass of beer. Coal bills from when he was in Vienna, month after month of utility bills, library slips—it was unbelievable."

Most unbelievable of all were these books filled with page after page of unintelligible writing. It was in no language that John Dawson had ever seen; even the alphabet was unfamiliar. There were these strange loopy figures and zigzags and dots. . . . What did they mean? Dawson didn't know.

But he was hooked. He decided he'd catalog Kurt Gödel's papers and personal records. Dawson was a scientist, and scientists are supposed to make discoveries. Maybe he'd decode these alien documents. Maybe he'd discover the answer to that so far undecided question, *Who was Kurt Gödel*?

Gödel was born on April 28, 1906, in what is now Brno, Czechoslovakia, and was baptized into a local German Lutheran congregation. He took religion seriously and apparently never lost his belief in God, something which unsettled his Institute colleagues. Gödel was supposed to have written a paper on the existence of God—a formalization of the so-called Ontological Argument—but some of the Institute regulars weren't looking forward to Dawson's finding it. "I think they were afraid it would somehow embarrass Gödel," Dawson says. "They were afraid that it was just evidence that he was off his rocker, or something like that."

Well, they had plenty of evidence of that already. Gödel was a hypochondriac even as a child. When he was about six or seven, he'd had a bout of rheumatic fever, which caused him a lot of pain and gave him the fright of his life. The disease often results in heart damage, and Kurt was convinced ever afterward that he had a bad heart, that he'd keel over and die if he didn't keep himself warm enough and eat just right. He was inquisitive, though, and asked all kinds of questions, about everything under the sun. *Herr Warum*, his parents called him, "Mr. Why."

In 1924 Gödel enrolled in the University of Vienna, where he intended to get a degree in physics. But he was impressed by the lectures of the number theorist, Philip Furtwängler, and changed his major to mathematics. Shy as a student, he was popular after class because he'd help anybody with their math homework.

Gödel's principal teacher, Hans Hahn, introduced his student to the famous Vienna Circle, a group of intellectuals including Hahn, economist Otto Neurath, and physicist Philipp Frank. Starting way back in 1907, these men used to meet on Thursday nights in local cafés for beer, cigars, and discussion, sharing their love of science and the scientific method. As their steins emptied and the smoke from their cigars rose into the rafters, they wondered why the scientific method couldn't be applied to everything, why indeed it couldn't be made to embrace the whole of human knowledge. By the time Gödel joined them in 1926, the group was meeting in a seminar room near the University of Vienna's mathematics department, where these young men were hammering out a new philosophy of science, a collection of doctrines that has since come to be known as "logical positivism." Based on the so-called verification principle, logical positivism holds that, in order to be meaningful, an assertion must be capable of being verified by sense experience. If you can't trace a concept back to something that you can see, hear, or touch, then the concept has no validity, i.e., it's meaningless. The first concept to go, naturally enough, was God, which was dismissed as just so much metaphysical blast and baggage.

Gödel was fascinated by all this talk—at least these people knew what the issues were, he thought—but he couldn't buy the party line. Already, at the age of nineteen, he had adopted his Platonist position in mathematics: he had decided that numbers and other mathematical entities were as real as anything else in the world. The fact that you couldn't *see* these things was of no consequence whatsoever. You couldn't see atoms either, but they were real enough. Later in life, Gödel claimed that this metaphysical Platonism—which he called his "objectivism"—helped lead him to his completeness theorem and other results in logic.

While he rejected the positivist ban on unseen entities, Gödel took an exceptional interest in the main scientific topic discussed at the Vienna Circle meetings, the foundations of mathematics. The problem was that the foundations were crumbling, and for a time the whole edifice of mathematics seemed to be tottering on the brink. Gödel wasn't much help here, though, and in fact, far from helping to rebuild the foundations, he sent the whole structure crashing headlong into the abyss, leaving other mathematicians—such as the Institute's Hermann Weyl—talking about "the Gödel debacle" and the Gödel "catastrophe." Indeed, what Kurt Gödel had to say about foundations was a shock to every working mathematician. He noticed that there were these slight cracks, these . . . *fissures* in the vault of the Platonic Heaven.

To the uninitiated, mathematics is the paradigm of certainty, a model of rational perfection and absolute truth. But mathematicians themselves soon learned the perils of thinking this way. From its very beginnings as a theoretical science, mathematics has had its share of inexactness and uncertainty. Indeed, some of the most basic mathematical facts remain mysterious even today. Take the case of "incommensurability," first discovered by the ancient Greeks.

Pythagoras—of Pythagorean theorem fame—was the head of a mystical society that claimed the world was composed of numbers, a belief which is not in fact as bizarre as it sounds. They viewed numbers spatially, as unit-points separated by distances. The number one is a single point; two is a pair of points which give rise to a line; three points make a triangle, four a square, and five a pyramid (see Figure 1).

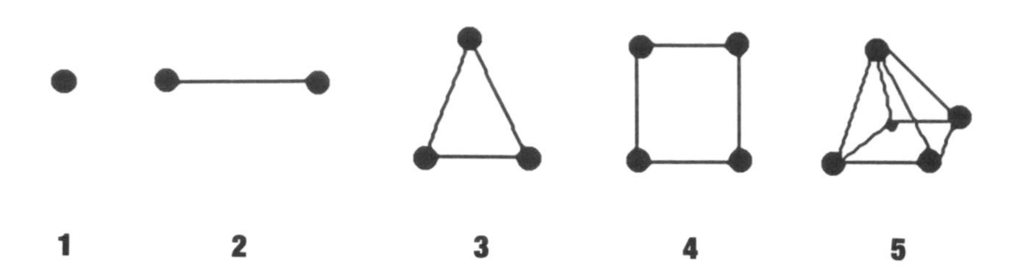

Figure 1 The Pythagorean integers

The Pythagoreans held the comforting view that mathematics is the supreme example of self-contained rationality, that it is lucid through and through, that it is, unlike the ordinary world of transience and decay, somehow a model of immutable perfection. What a surprise, then, for them to discover that, concealed inside the most symmetrical and four-square shape of them all, was something inexact, indeterminate, and in fact rationally incomprehensible. What they discovered was that the diagonal of a square is not measurable by the same units that measure the side.

If you take a ruler and measure the side of a square to be 1 foot long, then no matter how hard you try, no matter how finely divided the ruler's lines, you won't be able to get an exact measurement for the diagonal. The length of the diagonal will always fall somewhere *in between* the ruler's fine black lines—and not midway between any two of them, but somewhere off center. It's quite annoying. The same thing happens, though, if you try to determine the side's length arithmetically, using integers rather than physical measuring rods. Given a square with the side of 1, the Pythagorean theorem—$a^2 + b^2 = c^2$—tells us that the diagonal will be the square root of $1^2 + 1^2$, that is, $\sqrt{2}$. But $\sqrt{2}$ is 1.4142135 . . ., *and so on forever* (see Figure 2). In other words, $\sqrt{2}$ has no finitely expressible value: it's an "irrational" number, meaning that it can't be expressed as an integer or the ratio of two integers. That's pretty annoying, too, but to the Pythagoreans it was much more than annoying. It was a sign that there was something deeply wrong with the world, that there was a fundamental irrationality at the heart of nature.

As mathematical mysteries go, the problem of incommensurability

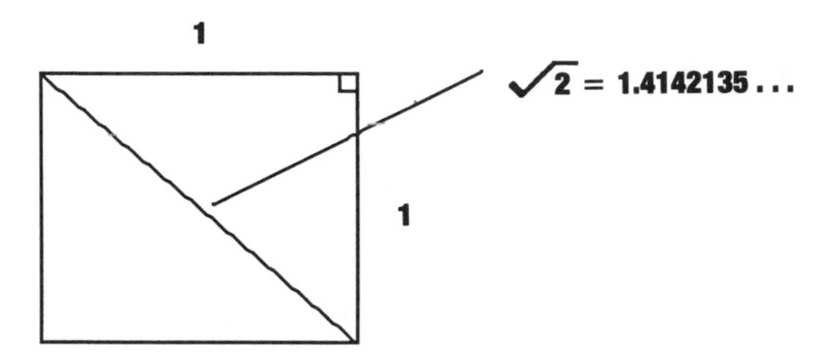

Figure 2 Incommensurability of the diagonal with the side of a square

is small enough; nothing really hangs upon it. But when analogous difficulties cropped up at the base of the calculus, mathematicians were faced with a crisis. Calculus is a method for dealing with "infinitesimals," very tiny stretches of time or distance. Suppose, for example, that you want to know the speed of a falling object as it passes a certain point x. Since its speed will be increasing at every point, it won't do simply to take an average of the speed at the beginning and at the end of its fall, for such a value almost certainly won't correspond to the exact speed at point x. You could progressively zero in on the correct value by taking an average of the speeds a little ahead of x and a little after x: that would give you a closer approximation, but still not the true answer. Leibniz and Newton, however, who co-invented the calculus, noticed that, if you narrowed down the interval *to practically nothing at all,* then you'd have the precise speed at x—which is exactly what you wanted to find. All you need is the ratio of an *infinitesimal distance* traveled, *ds,* divided by the *infinitesimal time* it took, *dt*. The true speed at point x will then be *ds/dt,* the derivative of *s* with respect to *t* (see Figure 3).

So far, so good, but what's all this "infinitesimal" stuff? Just how small *are* these infinitesimal intervals? And what's this business of narrowing down magnitudes *practically to nothing*? In his *Principia,* Newton called such infinitesimal magnitudes "evanescent" quantities, meaning that they disappear, they get smaller and smaller until—like in a magic show—they vanish completely. "By the ultimate ratio of evanescent quantities," he said, "is to be understood the ratio of the quantities not before they vanish, nor afterwards, but with which they vanish." In other words, infinitesimals are a ratio of quantities that in a sense . . . *aren't there.*

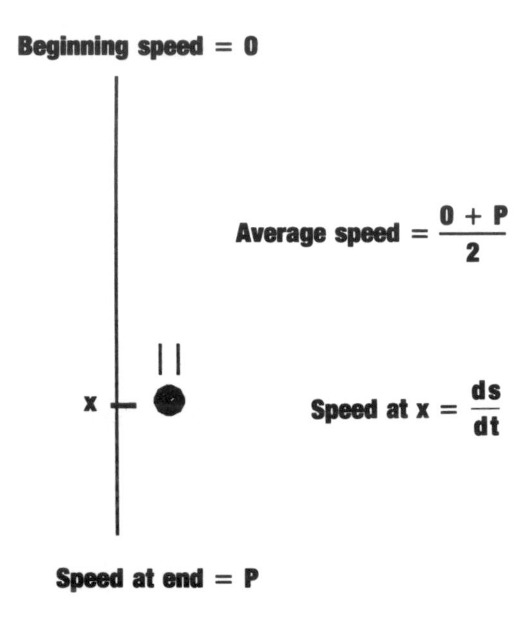

Figure 3 Calculating the speed of a falling object as it passes point x

Leibniz, for his part, sometimes spoke of infinitesimals as "imaginary," sometimes as "relative zeroes." But all this was pretty murky and mumbo-jumbo, and had philosopher George Berkeley delightedly quipping, "He who can digest [infinitesimals] . . . need not, methinks, be squeamish about any point in Divinity."

Infinitesimals were mysterious enough all by themselves, but they had their counterparts in the realm of the infinitely large. In the late nineteenth century, Georg Cantor was in the process of creating a new branch of mathematics called *set theory* when he came upon a paradox generated by the concept of the set of all sets. A set is simply any collection of objects. Integers can be regarded as objects, and so you can have a set such as {4, 7, 2, 3}, consisting of the numbers 4, 7, 2, and 3. This set, the same as any other set, contains within it many *subsets*, that is, sets equal to or smaller than itself. {4} is one subset, {4, 7} is another, {7, 3} is another, and so on. It turns out that a set containing four elements has sixteen possible subsets, and, in general, a set containing n members has 2^n subsets. Now the main thing to remember is that a given set contains many *more subsets than it contains elements*.

But how about the set of all sets? By definition it contains *all* sets as elements. But in view of the fact that a set contains more subsets than it contains elements, the conclusion follows that the set of all sets *contains more subsets than it contains sets*. Which is of course impossible.

By the time that Kurt Gödel was a student at the University of Vienna, mathematicians had gotten to the point where they wanted once and for all to banish such problems from the mathematical universe. As David Hilbert put it in 1925, "What we have experienced twice, first with the paradoxes of the infinitesimal calculus and then with the paradoxes of set theory, cannot happen a third time and will never happen again."

So mathematicians had to confront the general problem of "foundations," the task of providing a theoretical basis for their subject, a basis that would guarantee that, despite all the paradoxes, anomalies, and apparent impossibilities, mathematics was still the fount of certainty, the very epitome of absolute, proven truth that they had always hoped it was. The main thing to prove was that mathematics was at least *consistent*, that there were no contradictions hidden anywhere in its higher reaches. Another was to show that mathematics was *complete*, in the sense that it was capable of solving any problem that could arise within it.

Leader of the effort to reform the foundations of mathematics was David Hilbert himself, of the University of Göttingen. Renowned as the greatest mathematician of the day, Hilbert was an inveterate optimist. He was convinced that everything would turn out all right if you just worked hard enough at it, and he published articles and books describing how to revamp the foundations to eliminate all uncertainty. He was famous for his ringing pep talks, which he delivered at mathematical conferences. "Every mathematical problem can be solved," he declaimed. "We are all convinced of that. After all, one of the things that attracts us most when we apply ourselves to a mathematical problem is precisely that within us we always hear the call: here is the problem, search for the solution; you can find it by pure thought, for in mathematics there is no *ignorabimus* [We'll never know]."

Words to live by! A manifesto! The solution is out there—it's guaranteed to be—because *every mathematical problem can be solved* . . . because *in mathematics there is no ignorabimus*. Best of all, you can find the solution without experiments or apparatus, without consulting nature in any way. *You can find it by pure thought*. The mathematicians in the audience ate it up.

Hilbert, of course, had his own notions of how to go about proving all this, and he and his followers were soon making progress. In fact, they had almost licked the problem, they had almost shored up the crumbling walls of the Platonic Heaven . . . when absolutely the worst thing happened. As Gottlob Frege put it, "just as the building was completed, the foundation collapsed." Down there underneath it all, undermining the foundations at the very cornerstone, was none other than the quiet and reserved figure of . . . Kurt Gödel. He didn't strike with axe or sledgehammer. He didn't use a wrecking ball. It was more as if he had been tapping around for soundness with a little wooden mallet and had heard—*There it is!*—a slight hollowness within. And what a strange sound it was!

Gödel precipitated "the Gödel debacle" with his 1931 paper, "On formally undecidable propositions of *Principia Mathematica* and related systems." Here Gödel established that mathematics is not the all-encompassing, omnipotent system that Hilbert and others envisioned. He said in fact that the whole Hilbert project was hopeless, that there *are* some questions that arise within a mathematical system which the system itself is powerless to decide. In other words, there *is* an *ignorabimus* after all.

To prove his case, Gödel constructed just such an undecidable proposition within the *Principia Mathematica* system of logic. *Principia Mathematica* is an immense, three-volume work by the mathematical logicians Bertrand Russell and Alfred North Whitehead in which the authors hoped to show how all of mathematics could be derived from a few simple logical axioms and rules of inference. So all-encompassing and comprehensive was it supposed to be, said Gödel, that "one might therefore conjecture that these axioms and rules of inference are sufficient to decide *any* mathematical question that can at all be formally expressed in these systems. It will be shown . . . that this is not the case."

Gödel put together a proposition that, contrary to all expectation, *could not* be proven to be true in the system within which it was expressed. The clever part of it was that Gödel's undecidable proposition was the mathematical equivalent of the assertion, "This statement is unprovable." The proposition *was* in fact unprovable, but because of that very fact it was true: it was a true proposition that can't be proven to be true.

It sounded like the end of mathematics. It sounded as if, despite everything Hilbert had been saying over the years, and against all the instincts of the working mathematician, mathematics is inherently and inescapably imperfect and incomplete. It was depressing.

Sometime after Gödel's incompleteness paper was published in 1931, Oswald Veblen invited Gödel to come to the newly formed Institute for Advanced Study to give a course of lectures on his work. Gödel accepted, and on October 6, 1933, he arrived in New York. The Institute had opened its doors for the first time just four days earlier, with three faculty members: Alexander, von Neumann, and Veblen. Einstein was still in Europe.

The Institute was housed in Princeton University's Fine Hall, home of their mathematics department. This was a gabled, medieval fortress straight from out of the Dark Ages. From the outside it looked like a church, from the inside it looked like a dungeon. There were these stone-walled corridors leading off into dark-paneled rooms, all of them lit—if you could call it that—by these wrought-iron lighting fixtures that looked as if they were holding a dim candle or two. The place was gloomy. It gave you the feeling that you were inside the House of Usher.

In every way befitting the spirit of Fine Hall, the people who came there to pursue their advanced studies were called "workers." One of the Institute's first bulletins, dated February 1934, states that, "The workers are for the most part persons who received their doctor's degree some years ago, have engaged in university and college teaching, and have, while carrying on their routine, published papers indicative of promise." Under the heading "Workers Registered," the bulletin lists twenty-three people, seventeen from the United States, the rest—including one "Kurt Goedel"—from abroad.

Gödel lectured on his incompleteness results during the spring semester at the Institute, then traveled to New York and Washington to present his work to scientific groups there. In late May, he returned to Europe. Later, in the fall of 1934, at the age of twenty-eight, Gödel was admitted to the Sanatorium Westend, outside of Vienna, for treatment of nervous depression. Partly, Gödel was lovesick. He'd met a dancer, Adele Nimbursky, at a Viennese nightspot, and wanted to marry her. The problem was that Gödel's parents were opposed to the whole thing, no cabaret clogger for their little Kurtele. Gödel, always submissive to authority, obeyed his parents. It happened time and again throughout his career: Kurt Gödel, inestimable logician, fearless destroyer of mathematical perfection, he would prostrate himself before anyone in a position of institutional authority.

Deane Montgomery, who was Gödel's colleague at the Institute for many years, recollects a typical case. "Fifteen or so years ago we got into a big row with our director at that time, Carl Kaysen, about appointing somebody in the School of Social Science. Gödel read a lot of the papers of the man involved, and he thought they were terrible, really lousy—and that's what most of the faculty thought, and so we all voted against him. But Kaysen was determined to appoint this man, and the trustees supported Kaysen, so he was offered the job. It turned out that he didn't come here in the end. Gödel voted the same way as the rest of the faculty, but he said to me that there must be some explanation for the trustees' wanting to hire the man. 'Maybe they have some evidence about him that we don't have,' he said. 'Maybe this man has done some highly classified work for the government that we don't know about.' There were a number of incidents like that, where, even if he disagreed with authority, Gödel felt that in the end they must somehow be justified."

Another Institute mathematician, Atle Selberg, goes even further. "Gödel actually took literally the view that all authority is from God," he says. "He considered the law to be something sacred. He may not have been a practicing Christian in the sense of belonging to any particular denomination—he probably would have found some logical contradiction in their teachings—but he did believe in God and he clearly had a deeply held belief that this is where authority came from."

In the fall of 1935, about a year after he was treated in the Sanatorium Westend for nervous depression, Gödel returned to America for another stay at the Institute. Six weeks after he arrived, Gödel suddenly resigned from the Institute on account of overwork and depression, and two weeks later he departed for Austria. During the winter and spring of 1936, Gödel spent more time in sanatoria, and didn't resume any of his regular scholarly and teaching activities until over a year later, in the early summer of 1937, when he gave a course of lectures at the University of Vienna.

Two and a half years later, in the dead of January, 1940, Gödel and his wife—he had finally married Adele Nimbursky, the nightclub dancer—were hurtling across central Russia on the Trans-Siberian Railway. They were on their way to Yokohama, to board a ship for San Francisco, final destination, the Institute for Advanced Study.

In March of 1939, the Nazis had abolished university lectureships and created a new position called *Dozent neuer Ordnung* ("Lecturer of the New Order"). Gödel was out of a job, but as if that weren't enough he also received a notice to report to the Nazi army for a physical examination. So here was the shy, reclusive Kurt Gödel, at age thirty-two already the greatest logician of the twentieth century, married and jobless in a flat in Vienna, developing the distinct sinking feeling in the pit of his stomach that unless he did something fast he'd be out goose-stepping through the snow with the rest of the troops.

Hitler fever had already hit the city. When he got a bill from his cleaning lady, Maria Gabriel, Gödel found that it listed the total due, *Reichsmark 6.80*, and, underneath, a neatly typed "Heil Hitler!" Gödel didn't like any of this, so he started to go through the red tape that would get him and his wife to America. They'd decided that traveling by boat across the Atlantic was too risky, and so they'd take the long way around—across Russia, the Pacific, and the continent of North America. It was a long and tiring trip: they left Austria on January 18, and didn't arrive at San Francisco until March 4, 1940. Some days later, Gödel arrived in Princeton to begin his thirteen-year residency before being promoted to a professorship.

After his narrow escape from the Third Reich, Gödel must have experienced more vividly than many others the meaning of founder Abraham Flexner's description of the Institute as "a paradise," as "a haven where scholars and scientists could regard the world and its phenomena as their laboratory, without being carried off into the maelstrom of the immediate." By this time the Institute had moved from the medieval Fine Hall to the brand new Fuld Hall on the grounds of Olden Farm. When he moved into his second-floor office, almost directly above Einstein's, Gödel must have felt deeply relieved to be there. He had left the real world behind for good.

Or almost. In an account of Gödel's life and work, mathematician Solomon Feferman describes how Gödel almost did himself out of American citizenship. For this an oral examination was required, and so Gödel studied the United States Constitution. He noticed that, well . . . it had quite a few problems. For one thing, there were some contradictions in it. And for another, if you really looked closely enough, you'd find that the United States—quite legally!—could be turned into a dictatorship. He confided these discoveries to his friend, Oskar Morgenstern, who told him that he could not mention any of this at his citizenship examination.

On April 2, 1948, Gödel showed up at the government offices in Trenton, accompanied by Einstein and Morgenstern who were there as witnesses. On the drive down to Trenton, Einstein kept telling a bunch of stories and anecdotes to keep Gödel's mind off the logical problems of the American Constitution. But then the proceeding began. "Up to now you have held German citizenship, . . ." the official began, but Gödel jumped in and corrected this immediately. He was Austrian, not German. "Anyhow," the official continued, "it was under an evil dictatorship, but fortunately that's not possible in America. . . ."

"On the contrary," Gödel cried out, "I know how that can happen!" Finally, though, Einstein and Morgenstern succeeded in restraining Gödel long enough for him to be examined and duly sworn in as a citizen of the United States.

Gödel was never one for tactfulness or the social niceties. Once Hermann Weyl—the mathematician who spoke of "the Gödel debacle"—invited the mathematician Paul Lorenzen to come for a stay at the Institute. Lorenzen had written papers that tried to patch up the damage done by Gödel's bombshell, and these were all very much to Weyl's liking. "I see the heavens open again," Weyl wrote to Lorenzen.

Unfortunately, Weyl died before Lorenzen got to Princeton, and so when he showed up at the Institute he was greeted by Gödel. Practically the first thing out of Gödel's mouth was, "I know your works—and regard them as harmful." This made quite an impression on poor Lorenzen: "You will understand that I remember this sentence in its exact wording ever since," he says.

On another occasion, Gödel attended one of the faculty dinners that the Institute used to throw once or twice a semester, and sat across from John Bahcall, a rising young astrophysicist. The two introduced themselves, and Bahcall said he was a physicist, to which Gödel replied: "I don't believe in natural science." Of course not: natural science is messy and inexact; it doesn't reflect the immutability of the true mathematical objects. Bahcall understood all this, but still, conversation that night was a little difficult.

Gödel's main scientific project at the Institute for Advanced Study was an examination of the continuum problem. As Gödel himself described it, "Cantor's continuum problem is simply the question: How many points are there on a straight line in euclidean space? An equivalent question is: How many different sets of integers do there exist? This question, of course, could arise only after the concept of 'number' had been extended to infinite sets."

Prior to Cantor's time, and in fact going back to the days of the Greeks, mathematicians denied the existence of any sort of infinite totality. The infinite could exist, but only potentially—as a *process*—meaning that you could always produce a larger number by adding 1 to any given number. Since you would never run out of 1's, this process could continue on endlessly, and so the numbers were in that sense "infinite." Nevertheless, you'd never have any *specific* infinite number in front of you at any time, for any particular number was by definition finite.

Cantor, though, disagreed with this reasoning. "I place myself in a certain opposition to widespread views on the nature of the mathematical infinite," he said. This was an understatement. He ended up postulating not one, but a whole series of infinities, an endless hierarchy of them, in fact. While some mathematicians viewed this as sheer insanity, others thought that Cantor had opened up a marvelous new mathematical universe. Hilbert, for one, spoke of "the paradise which Cantor created for us."

As far as Cantor was concerned, there was no reason at all why you couldn't have an entire infinite set right there in front of you. After all, mathematicians made use of infinite quantities all the time, although they might not have noticed the fact. Take pi (π), for example. It had been known since 1767 that π is an irrational number, meaning that there is no finite decimal expansion which gives its exact value. Nevertheless, pi *does exist*: it's simply the ratio of a circle's circumference to its diameter, and so if circles exist, and if their diameters exist, then so must pi. The fact that you could never fully *write out* all pi's digits doesn't matter. Pi exists, although it continues on endlessly.

But Cantor went much further. He said that you could have infinite sets of *different sizes*. The set of natural numbers, that is, the counting integers 1, 2, 3, and so on, was only the *smallest* of these sets. On top of that there was a whole class of larger infinite sets, a collection of bigger and bigger infinities.

Now, to keep track of all this, Cantor decided to designate infinite sets by means of his own special symbols, based on the first letter of the Hebrew alphabet, aleph: $\aleph$. To the smallest infinite set, the set of natural numbers, Cantor assigned the cardinal number aleph-null, or $\aleph_0$. Cantor said that two sets were equivalent if they could be put in a one-to-one correspondence with each other. So, since the set of unit fractions can be put into a one-to-one correspondence with the set of counting numbers, it follows that the two sets are equivalent in number, both having $\aleph_0$ items:

1	2	3	4	5	...
↕	↕	↕	↕	↕	
1/1	1/2	1/3	1/4	1/5	...

But far larger infinite sets exist. For example, between any two whole numbers, say, between 1 and 2, there's a whole other infinity of *decimal* numbers. There's .25 and .26, and between them there's .255 and .256, and between *them* there's .2555 and .2556, and so on forever. In fact, between *any* two decimal numbers there are an infinite number of smaller decimal numbers. What this means is that the decimal numbers—the whole lot of them, regarding even whole numbers, such as 1.00, as decimals—constitute an infinite set with far more members than the set of whole numbers itself.

Now this may seem like sleight of hand. After all, if you have an endless supply of integers, then surely there are enough of them to be put into a one-to-one correspondence with the decimal numbers. It seems as if there *must* be enough, if you can't run out of them.

But Cantor had a proof that this is not so: he proved that there are far too many decimal numbers to be put into any such one-to-one correspondence with the set of counting integers. He proved this with his famous "diagonal argument."

Consider the following simple arrangement of numbers:

1 2 3
4 **5** 6
7 8 **9**

The number formed by the diagonal running from top left to bottom right gives a new number, **1 5 9**, which—in this case, anyway—is not on the original list. But Cantor discovered that for any arbitrarily large array of numbers there was a way of constructing a new diagonal number that was not on the original list. Using this method, he showed that there are in fact more real numbers than there are counting numbers.

Suppose that the opposite were true, that is to say, suppose that the real numbers *could* be put into a one-to-one correspondence with the counting numbers. Confining the argument to just the real numbers between 0 and 1, we might start matching up counting numbers and real numbers like this:

Counting numbers	Real numbers
1	.**0**000000001234 . . .
2	.3**4**75869979787 . . .
3	.98**8**4666567576 . . .
4	.757**2**574543298 . . .
5	.6666**6**66666666 . . .
6	.02988**4**7244656 . . .
7	.500000**0**000000 . . .
.	. . . .
.	. . . .
.	. . . .

Imagine that the list continues on infinitely, throughout the whole series of integers. Since the integers never give out, it seems that there ought to be enough of them to allow the listing of every last real number, no matter how many there are. But as the diagonal argument shows, this is a false assumption. Take the number formed by the diagonal, namely

.0482640 . . . ,

and *alter each digit,* simply by adding 1 to each of them. Doing this gives us a new number, **n**:

n .1593751 . . .

Now it turns out that this new number **n** will not be found anywhere on the list of real numbers. The reason is that **n** *will differ by at least one digit from any number already on the list*. Note, for example, that

- the first digit of **n** differs from the first digit of the first real number; therefore **n** is different from it;
- the second digit of **n** differs from the second digit of the second real number; therefore **n** is different from it;

- the third digit of **n** differs from the third digit of the third real number; therefore **n** is different from it,
- and so on.

In other words, if **n** is different from every listed real number, then **n** is *not* in fact on the list that supposedly contained "all" the real numbers. The upshot is that, contrary to the original supposition, you *can't* set up a one-to-one correspondence between the integers and the reals: there are just too many more reals than there are integers to number them.

Now the continuum problem is the question whether there is an infinite set *larger* than the set of natural numbers (the integers), but *smaller* than the set of real numbers. It's a simple enough question, and it sounds as if it ought to be easy enough to answer. Cantor himself thought that the answer was no, that there was *no* infinite set in between the "small" infinite set of integers and the "large" infinite set of decimals. This view, that there is *no* such intermediate set, is nowadays referred to as *the continuum hypothesis*. The only trouble with the continuum hypothesis is that Cantor couldn't prove it—and neither could any other mathematician.

At the Institute for Advanced Study, Kurt Gödel spent many of his years trying to solve the continuum problem. Like Cantor, he never solved it either, but he did make some progress. He formulated a mathematical model in which the continuum hypothesis held: he showed that given the axioms of "standard set theory"—as stated by Ernst Zermelo and Abraham Fraenkel—the continuum hypothesis could be attached as an additional axiom without any contradiction arising as a result.

Later, Gödel took on as an assistant the young mathematician Paul Cohen. Cohen and Gödel worked on the continuum problem together for two years, and, shortly after he left the Institute, Cohen made an additional discovery. Using a novel set-theoretic strategy he developed called the "forcing" method, Cohen established the result that the continuum hypothesis was *independent* of the Zermelo-Fraenkel axioms. Just as the parallel postulate is independent of the other axioms of euclidean geometry—meaning that it is not provable by reference to them—Cohen showed that the continuum hypothesis could not be proven only on the basis of the axioms of set theory. On his former student's behalf, Gödel submitted Cohen's proof to the *Proceedings of the National Academy of Sciences*, where it was published in 1963. But the continuum problem is still there today, waiting for a full and final solution, a chestnut in theoretical mathematics.

Gödel's best friend at the Institute was Albert Einstein, and it was a common sight to see the two of them walking to the Institute together, the two solitary Olympian masters of their respective disciplines. Neither of them tended to pal around with the others much. Every once in a while the Institute mathematicians would have lunch in town—at the Nassau Inn—and Gödel would come along and loosen up a bit, maybe crack a joke or two. Most of the time, though, Gödel would sit by himself in the Institute dining hall and have a cup of tea and maybe eat an apple. In an Institute full of oddballs, Gödel was still an odd man out.

At some point, Einstein evidently managed to interest Gödel—hitherto a mathematical logician of the purest stripe—in physics, particularly in general relativity theory. One day Einstein happened to meet Gödel in the mathematics library in Fuld Hall, and here they discussed certain problems in solving Einstein's gravitational field equations. Gödel worked on the equations afterward, and produced some new solutions, according to which the lapse of time, and therefore the existence of change in the natural world, could be viewed as unreal and illusory. "It seems," Gödel said, "that one obtains an unequivocal proof for the view of those philosophers who, like Parmenides and Kant, and the modern idealists, deny the objectivity of change and consider change as an illusion or an appearance due to our special mode of perception."

Gödel published the results of his work on general relativity in a 1949 paper titled "An Example of a New Type of Cosmological Solutions of Einstein's Field Equations of Gravitation." His solutions describe a "rotating universe," a world in which matter everywhere revolves as if in a vast, cosmic whirlpool. This rotation of matter gives rise to space-time trajectories that loop back upon themselves, returning to places they've already been. It follows from this that time is not a straight linear sequence of events, but something that bends around the universe in a curving line. You could travel from one point on the curve to another, Gödel thought, if you had a fast enough spacecraft. "By making a round trip on a rocket ship in a sufficiently wide curve, it is possible in these worlds to travel into any region of the past, present, and future, and back again."

So Kurt Gödel would walk through the streets of Princeton thinking that change is an illusion . . . that infinite sets of numbers exist off in the Platonic Heaven . . . and that you can go backwards and forwards in time. With theories like these, of course, he was fully qualified to be an Institute professor.

In 1953, after being at the Institute continuously for thirteen years, Gödel was promoted to professor of mathematics. Some of the faculty were incensed at the delay. "How can any of us be called professor when Gödel is not?" John von Neumann asked. Why it took so long has never been explained. Some say that the administration wanted to do Gödel a favor, that they wanted to avoid subjecting him to the responsibilities that would come with professorship—the obligation to take part in faculty governance, to review applications for promotion, temporary memberships, and so on. The other view is that Gödel, with his awe for authority, with his legalistic mind and generally nutty approaches to things, would interfere with the conduct of Institute business. Perhaps—God forbid!—he'd find a contradiction in the Institute bylaws.

As it turned out, Gödel took his new faculty responsibilities quite seriously, and even seemed to cherish the extra work. He would scrutinize applications for membership endlessly. "It was hard to get Gödel to say something about these people," says Atle Selberg. "And often very difficult to get their files out of his hands."

In later years Gödel moved from his office in Fuld Hall to the brand-new Historical Studies library. This is a fabulous building, winner of architectural awards, glass and concrete everywhere. But if Gödel was out of touch beforehand, he was even more out of it now. The library is at the far end of the campus, and Gödel's new office had large floor-to-ceiling panes of glass that gave him a panoramic view of a small pond, stately trees, and the woods farther off. The view brings a feeling of utter serenity and peace, of being at one with nature. The outside world doesn't seem to exist here at all. There's no noise, no people, no conflict. There's only silence, Gödel, and numbers.

By the time he moved into his new office, Gödel was just about finished as a publishing mathematician. He revised his earlier papers, continued to work on the continuum hypothesis, and plunged into philosophy, studying the writings of Leibniz and Edmund Husserl. "It is strange how little he published," says Paul Erdös, a Hungarian mathematician who used to visit Gödel at the Institute. "I always argued with him. We studied Leibniz a great deal and I told him 'You became a mathematician so that people should study you, not that you should study Leibniz.'" But there were no obligations at the Institute—"no duties, only opportunities," said Flexner—and Gödel, like everyone else there, did what he wanted. He'd emerge from his cocoon now and then to accept honors and awards, but just as often would decline invitations, for reasons of "ill health." At his home on Linden Lane he enjoyed reading about theology and religion—and about ghosts and demons.

Always preoccupied with his ailments, real or imagined, and a notoriously abstemious eater, Gödel became increasingly wan, withered, and frail. He experienced new bouts of depression, for which he twice consulted Princeton psychiatrist Philip Erlich. At one point he was on the verge of admitting himself into a psychiatric hospital, and asked the advice of some of the other mathematicians on the faculty. One told him to go, another told him not to, and in the end he never went.

Gödel did go to see Dr. W. J. Tate, of the Princeton Medical Group, in February of 1970. He told Dr. Tate that he wanted to have an EKG to prove that he needed digitalis as a tonic for his heart. He claimed that he'd been under the care of another physician in town who prescribed digitoxin, but Gödel thought that digitoxin was *toxic*, that it was poison. Tate disabused him of this notion—or at least tried to.

Later that same month, Gödel made an appointment to see Dr. Harvey Rothberg, a Princeton internist, for a one-hour consultation. But instead of coming in for the visit, he made two phone calls to discuss at great length with Rothberg's nurse the tests he'd be undergoing. Finally he called back a third time to cancel the appointment "because," Rothberg says, "he had heard that it was cold in my office." Later on Gödel called back to apologize. He made another appointment, but he broke that one, too.

"After breaking two appointments," Dr. Rothberg says, "and [after] approximately fifteen telephone calls, the patient finally appeared one day at 5 P.M.—which is the time the office closes—without an appointment, to discuss his problem."

Rothberg nevertheless saw Gödel. The medical records show that, although he was 5′ 6″ tall, Gödel weighed only 86 pounds.

"His thought content was somewhat paranoid," Rothberg says, "and he had fixed ideas regarding his illness, and my diagnostic impressions were that there was indeed a severe personality disturbance, with secondary malnutrition, and some somatic delusions—inappropriate ideas about his bodily structure and function." Rothberg suggested that Gödel take some additional vitamins, and that he might be helped by a psychotropic drug.

Gödel made another appointment for a return visit about two weeks later, but then canceled it. "When he did come in several days after that," Rothberg says, "he began the conversation by asking whether I was the real Dr. Rothberg. I'm not sure why he was in doubt of that. He told me that he had decided to see a psychiatrist, and that he had an appointment for that purpose, and he inquired whether there was a law against mercy killing."

In 1974 Gödel was hospitalized for a urinary tract problem related to his prostate and was advised by at least two urologists, Dr. James Varney and Dr. Charles Place, to have an operation. Even though his own father had died from a prostate condition, Gödel rejected surgery.

On July 1, 1976, at the age of seventy, Gödel retired from the Institute for Advanced Study. He'd been there continuously for thirty-six years, and had been a visiting member before that. He first arrived there about the same time as Einstein and von Neumann and the other founding fathers, and he went on to outlive them all. He was an emblem of the place, but the most vivid memory that Institute old-timers have of him is that of a cadaverous old man shuffling past alone, dressed in his black coat and winter hat.

At the end of it all, Gödel was depressed. He thought that he had let the Institute down, that he had not achieved enough. Stanislaw Ulam had the impression that, despite the fame brought to him by his undecidability theorem, Gödel suffered from "a gnawing uncertainty that maybe all he had discovered was another paradox à la Burali Forte or Russell."

A year after his retirement, Gödel's wife Adele underwent major surgery and for a while was put in a nursing home. This was the kiss of death for Kurt Gödel. Adele had been nurturing her husband all along, getting him to eat when he otherwise wouldn't touch a bite of food—because it was poisoned. With his wife in the hospital, Gödel had to fare for himself. He responded by not eating at all, and he progressively starved himself to death.

On December 29, 1977, Hassler Whitney, Gödel's colleague from the Institute, called Dr. Rothberg from Gödel's home, saying that the patient was dehydrated and not doing well. Whitney brought Gödel to the emergency room of Princeton Hospital, where he was admitted.

Gödel persistently refused to eat and, after two weeks in the hospital, died sitting up in his chair on the afternoon of January 14, 1978. According to his death certificate, Kurt Gödel died of "malnutrition and inanition caused by personality disturbance." His wife Adele died three years later, and both are buried in Princeton Cemetery. Kurt and Adele had no children, and Kurt's brother Rudolf, a retired radiologist in Vienna, never married, and so the Gödel family line is thus without further issue.

It took John Dawson two years to go through all sixty boxes of Gödel's personal papers and documents. He sifted through and catalogued every last scrap of them, and now any scholar can go to Princeton and, piece by piece, trace through the flotsam and jetsam of the life and work of Kurt Gödel. Here, in folder after folder—catalogued, stamped, numbered—are Gödel's bankbooks, canceled checks, passports, electric bills. Here is the homework—sheets and sheets of truth tables—done by one of Gödel's students, an F. P. Jenks, in May 1934, when Gödel was teaching a logic course at Notre Dame. Here's a picture of Gödel in pajamas and bathrobe, sitting in his living room.

You can see a bookseller's statement, dated 21 July 1928, for Gödel's copy of Whitehead and Russell's *Principia Mathematica*. Did Gödel—then twenty-two years old and three years away from publishing his blockbuster paper on the "undecidable propositions of *Principia Mathematica* and related systems"—did he have a glimmer yet of the havoc he was to cause with the book he was buying?

There are Institute for Advanced Study salary statements, tiny, neatly kept expense ledgers, and the books filled with the baffling, alien, loopy script. Dawson finally figured out what these are. They're Gödel's notes on various subjects, written in an obsolete German shorthand called "Gabelsberger," named after Franz Xaver Gabelsberger, its inventor. Later, Dawson discovered a "Rosetta stone" in the form of Gödel's shorthand textbook and practice workbooks, and even located an expert on Gabelsberger in New York City, a photographer by the name of Hermann Landshoff. It will be a while before all Gödel's notes are decoded, though. There are lots of them, and Gödel made some of his own alterations to the shorthand.

There's a picture of Gödel in a swimming pool, and a shot of him drinking a glass of beer. Sometimes he's even smiling. And there are, finally, folder upon folder of letters and manuscripts from other mathematicians, from logicians the world over, cover letters sent together with journal offprints proving some little theorem or other, letters from every assistant professor of mathematics in the country, hoping that Gödel's eyes should gaze upon the results of their work, hoping for a nod of recognition, a word from the master, the merest sign of approval from the high lonesome God almighty of logic.

Chapter 4

Behold the Forms

Is this a dagger which I see before me? . . . God knows *what* it is.

John Milnor is staring at his computer display screen, peering deep down into it, as if he could see below the glass surface, into the very insides of the thing, right down to the electrons themselves as they stream off the emitting element, the hot cathode in back. Milnor is the prototypical Institute professor, tall and thin, way up there on the Gary Cooper scale. His long grey hair sometimes swings down in front of his eyes as he gazes into the screen.

He's sitting in front of the computer, squinting at this weird black *shape*. It seems to be a picture of . . . *something* . . . but of *what* in the world? Whatever it is, it's diseased-looking, bristling with humps, bumps, and all manner of erratic pods and appendages. They look like flower stamens, perhaps, or a map of the dendritic network of the brain. Or is it one of those "irregular" galaxies or anomalous stellar clusters that you see in astronomy books? Whatever it is, it's truly . . . alien, not recognizable at all

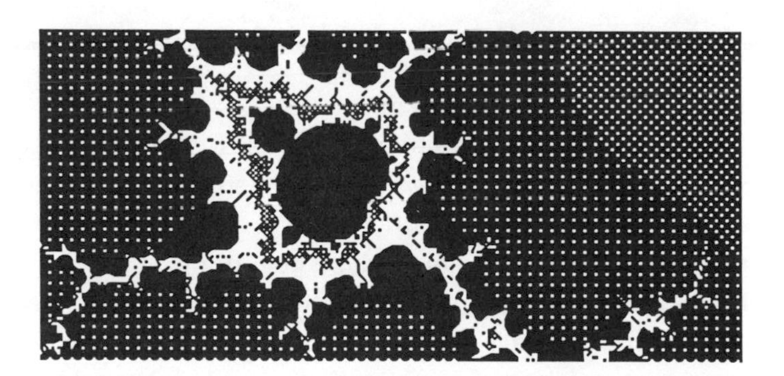

The fact is that John Milnor is not a botanist, a neurophysiologist, or an astronomer. He's a mathematician, and a pure mathematician at that, entirely theoretical, not one of those applied math-engineering types who spend their days solving practical real-world problems—like figuring out the loads on a suspension bridge. No, John Milnor is a true Institute man, head-in-the-clouds, and all the rest of it. But then . . . what in the name of God is he looking at, what is this bizarre unearthly *thing* quivering there in black-and-white phosphorescence?

Milnor types something into the computer keyboard and the image seems to rise, as if it's coming toward you. The form begins to enlarge, looming up at you now, its blackness threatening to engulf the whole screen. As it gets bigger you can see that one of the . . . pods, whatever it is . . . has a filament attached to it. It goes off in a zigzag, like a bolt of lightning. The whole picture gets even bigger and then . . . then you can make out some even smaller, finer filaments, only they're chainlike, broken off at points. You have to strain to see them. They're at the threshold of visibility, right at the limit of the computer's ability to represent a line.

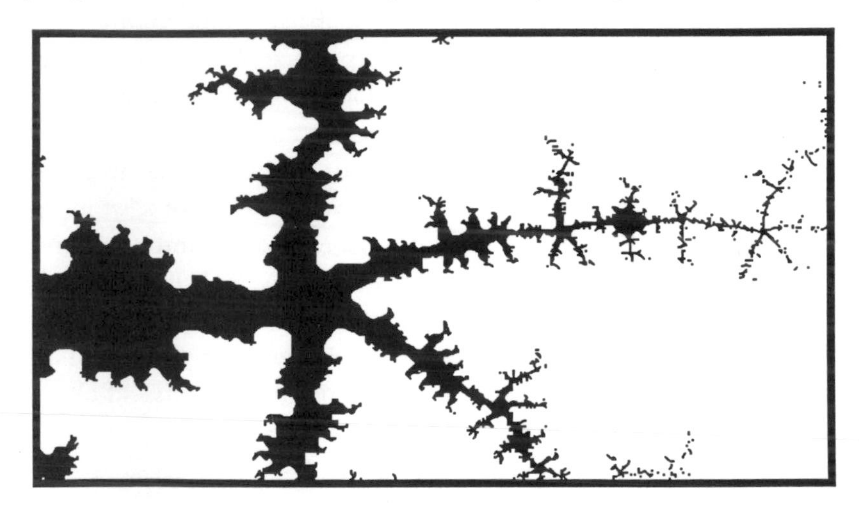

Milnor points to one of the chains.

"You see that little squiggle there?" he asks. "Well if you continued to magnify it you'd find that there would be still more of these little filaments in there wandering in and out. What I'm trying to do is to understand the structure of it, the behavior of the whole thing."

Well, that's a great idea, very worthy. I'd like to understand it myself, that's for sure. Because what you have there on the screen *is* a picture . . . but it's not a picture of anything that exists in this world. And it's not just a "computer graphic," one of those fake landscapes or computer-generated still-lifes that you see every once in a while in the glossy high-tech magazines. No, this is something *real*, only it's not of *this* world. The fact is that the shape hanging there, floating in electronic space on the display screen, is a mathematical object, a pure abstraction made visible.

It's like having a glimpse of the Forms themselves, a direct vision of the Platonic Heaven.

Twenty-five years ago no pure mathematician would have recognized the image on Milnor's screen. Twenty-five years ago, in fact, that object did not exist. Nowadays, though, it's regarded as "the most complex object in mathematics."

Back when the Institute first opened, theoretical mathematicians concerned themselves with clean, precise entities—things like numbers, the transcendentals, for example, or the functions of a complex variable. Even today, number theory is big business at the Institute for Advanced Study. Not just any numbers will do, of course, there has to be some rhyme or reason among them, an *order* to it, as there is among the primes.

Prime numbers are those evenly divisible only by themselves and the number 1. The number 5 is a prime, for example, because it can't be divided without remainder by anything other than 1 and itself. There are a lot of primes, and it's an easy matter to list the first few of them:

2 3 5 7 11 13 17 19 23
29 31 37 41 43 47 53 59 61
67 71 73 79 83 89 97 101 . . .

But so what? They're just a lot of numbers, why prime number *theory?*

The fact is that the prime numbers constitute no more and no less than the scaffolding of the integers. They're the backbone of the number system in much the same way that the chemical elements are the foundation of chemistry. Just as any compound substance is composed of one or more of the chemical elements, all non-prime numbers are *composed* of prime numbers. The primes are the units, the irreducible primordial constituents, of all the rest of the whole numbers, a fact which has become enshrined as *the fundamental theorem of arithmetic,* which states that every integer greater than 1 can be factored into a unique set of primes. The number 12, for example, is the product of the prime numbers $2 \times 2 \times 3$, whereas it is not the product of any other set of primes. Likewise, 100 is factorable into the four prime numbers $2 \times 2 \times 5 \times 5$, but not into any others. So the primes are the building blocks, the elemental constituents of everything else in the numerical universe.

But that's only the beginning. Part of the lure of the prime numbers is that, although they're simple, plentiful, and easily producible, further knowledge about them is exceedingly hard to come by. Is there a largest prime number? Is there a formula for producing all the primes? Is every even number the sum of two prime numbers? After centuries of theoretical work by pure mathematicians, only one of these questions is known to have a definite and provable answer. Euclid proved that there is no largest prime number.

To a nonmathematician, Euclid's proof may seem gratuitous: why should anyone think that there *was* a largest prime? The answer is that the primes get rarer and rarer the higher up you go in the series of natural numbers. The first four primes (2, 3, 5, and 7) are separated by at most a single integer, but as the numbers get larger, the gaps between successive primes increase. After 811, for example, there is not another prime until 821, ten integers away. With still bigger numbers the gaps stretch out even more, so that in the hundred integers between 10,000,000 and 10,000,100 there are only *two* prime numbers: 10,000,019 and 10,000,079. Farther out, among the colossally large numbers, there's a veritable prime desert, a stretch of one million integers without a single prime number among them.

In retrospect, the increased rarity of primes makes perfect sense. The larger the number, the more numerous the smaller integers that could divide it evenly. It was this fact that got some mathematicians to thinking that perhaps there's a point somewhere—way up in the higher reaches of the number series—at which there are so many smaller integers that a new prime number just doesn't exist. Maybe, sooner or later, the primes just *run out*.

Euclid, however, proved that the primes go on forever. He employed a strategy that number theorists still use today, called proof by contradiction. In essence, you assume the opposite of what you want to prove, and then you derive a contradiction from that assumption. Since contradictions aren't possible, it follows that the assumption which generated the contradiction must be false, and, therefore, that its opposite—the statement that you had originally set out to prove—must be *true*.

Euclid started out by assuming that there *was* a largest prime number. Suppose, for example that the largest prime number was 5. (Suppose.) If that were true, then the set {2, 3, 5} would be the set of *all* the prime numbers, and so it would be impossible to produce a bigger one. But of course there *is* a bigger one: all you have to do is to multiply together the elements of the set {2, 3, 5}, then add 1 to the result. In other words, you take $(2 \times 3 \times 5)$, which gives you 30, then you add 1, which gives you 31. Now 31 is a prime number, because whatever you divide it by, there is always a remainder. So, contrary to the assumption that 5 is the largest prime, you have constructed a prime that is even larger.

Now the nice thing about this reasoning is that it's reiterable, that is to say, it can be used again and again on *any* set of primes, as large as you please. So if 7 were the largest prime, then you could make an ever bigger prime number by multiplying together "all" the primes $(2 \times 3 \times 5 \times 7)$, then adding 1 to the result. This produces the number 211, which is a prime.

On the rarefied level of number theory you'd express all this abstractly:

To prove: There is no largest prime number.
Assumption: The largest prime number is *n*.
Proof: The assumption can't be true, because if it were, then the set {a, b, c, . . . *n*} would be the set of *all* the prime numbers. However, multiplying together all the elements of that set and adding 1 to the result, that is, taking

$$(a \times b \times c \times \ldots \times n) + 1,$$

will give us some new number **t**. But **t** is not evenly divisible by any member of the set of primes {a, b, c, . . . *n*}, since dividing **t** by any element of that set will always leave a remainder of 1. So either **t** itself is a prime, or there must be another prime p, larger than *n*, which divides **t**. In either case, *n* will not be the largest prime number. Which means that there is no largest prime.

Some other claims regarding the primes turned out to be horrendously difficult to prove or disprove. The claim that every even number is the sum of two primes has never been proven, and has been known as "Goldbach's Conjecture," since Christian Goldbach first made it in 1742. Likewise, a recipe for producing all the prime numbers has never been found. Some formulas, it's true, are successful in producing long strings of primes. Leonhard Euler's formula $x^2 + x + 41 = p$, for example, yields 40 prime numbers, one right after the other. (For example: When $x = 1$, $p = 43$, which is a prime; when $x = 2$, $p = 47$, which is also a prime.) Nevertheless, of the first two thousand numbers produced by Euler's prime-number generator, only about half of them turned out to be genuine prime numbers.

In the spring of 1948, as he was finishing up a one-year temporary membership at the Institute for Advanced Study, Atle Selberg, thirty-one, was making some progress toward a proof of another such conjecture, known as *the prime-number theorem*. Selberg was from Norway—he had gotten his Ph.D. from the University of Oslo in 1943—and was shortly to take up a position at Syracuse University, in New York. But in the summer of 1948, while he was still at the Institute, Selberg discovered the final step needed to complete his proof. It was a major accomplishment, because the prime-number theorem had been lying around without a proof since the time of the mathematicians Karl Friedrich Gauss and Adrien Marie Legendre who, around the year 1800, and working independently of each other, noticed a relationship between the density of prime numbers—that is, their frequency of occurrence as you ascend through the integers—and a logarithmic function that governs exponential growth and decay. Gauss and Legendre observed that, as you moved upward through the integers, the number of primes tended to converge toward the number predicted by the logarithmic function.

Integers (x)	Number of primes	Number predicted by x/log x	Error
1000	168	145	16.0%
1,000,000	78,498	72,382	8.4%
1,000,000,000	50,847,478	48,254,942	5.4%

Simply put, the prime-number theorem states that these rates will converge toward a limit, that the difference between them will approach zero. This was a totally unexpected coincidence, for the two numerical progressions seemed to be essentially unrelated: on the one side of the table are the prime numbers in all of their unpredictability and apparent randomness; on the other, a continuous logarithmic function coming right out of the calculus. Evidently there was some deep and unseen connection here somewhere, but the question was, is the connection genuine or accidental? And, above all, can any such connection be proven?

Jacques Hadamard and C. de là Vallée Poussin advanced proofs of the prime-number theorem in 1896, but their proofs were fantastically difficult, involving complex variables and wave analysis, things so much more complicated than the prime numbers themselves that mathematicians felt that there must be a simpler proof somewhere.

It was just such a proof that Atle Selberg found in the summer of 1948. He wrote up his argument, giving it the title of "An Elementary Proof of the Prime-Number Theorem," and published it in the *Annals of Mathematics* the following year. Selberg's proof is anything but simple. Number theorist Ian Richards observes that the proof may be "technically 'elementary' in the sense that each step is elementary. But there are so many steps, and the way they fit together is so complicated, that no simple picture emerges." The Institute's mathematics faculty, meanwhile, wanted to keep Selberg at the Institute, but there was problem as to rank: should Selberg be given a permanent membership or a full professorship?

The question pointed to the Achilles' heel of the Institute for Advanced Study. You don't want to hire people as professors unless they've made crucially important contributions to their discipline, but on the other hand you don't want to hire a person who has seen his best days. In the sciences, however, and especially in mathematics, it's a truism that once a person has done a really important piece of work, then that person's best days are over. There was not the slightest doubt that Selberg's proof was important work, but the worry was that, if brought in as a professor, he might turn into a fossil in his early thirties. Oppenheimer, who was director at the time, accordingly suggested that Selberg be awarded a permanent membership, a comfortable position by any standard, but still something less than a full professorship. (In those days a permanent membership went for $9,000, a professorship for about $15,000.) Oswald Veblen, though, wanted Selberg to come in as a professor.

The Institute's faculty and board of trustees agreed with Oppenheimer, and Selberg was made a permanent member. A year later, however, he had received the Fields medal, the highest honor in the discipline (there is no Nobel prize in mathematics), and the Institute was forced to reexamine its decision. In the spring of 1951, Atle Selberg was made a full professor, a position he holds today. During the interim, Selberg has continued to do mathematics, although it's an open question whether he's done anything as important as his proof of the prime-number theorem. His own feeling, at any rate, is that he did his best work after he became an Institute professor, when he worked on a complex problem called the trace formula.

John Milnor moves to another portion of the object on the screen. He wants to show me its delicacy, its true beauty. He studies the object not because it's useful, he says, but because it's pleasing to the eye. "My motivations are primarily aesthetic. I look at these things because they're beautiful in their own right."

The shape trembles a bit, then it starts to move down the screen, so that more of the top of the object comes into view. Finally, a magnified portion of it appears and stabilizes. It looks like nothing so much as the tracery of riverbeds, or an overhead view of the Grand Canyon. Or maybe it's a portion of the lungs' alveolar system.

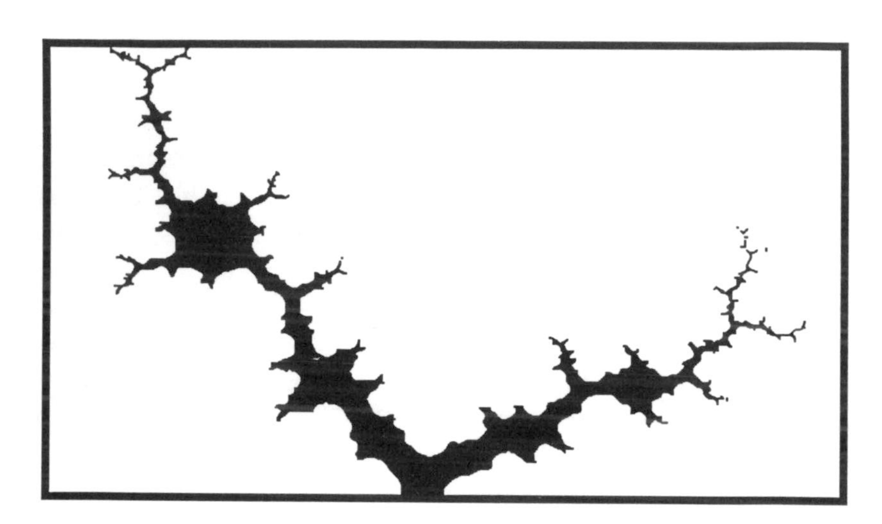

Pure mathematics is a serious business, and its practitioners some of the more sober people at the Institute, but when in 1958 the School of Mathematics hired André Weil as a professor, they were taking on one of the greatest confabulators and practical jokesters in the history of the discipline. Weil is one of the perpetrators of the "Nicolas Bourbaki" hoax.

Nicolas Bourbaki is the author of some two dozen books of mathematics, a series which commenced in the late 1930s with *Éléments de Mathématique*. Billed as a revolutionary synthesis of contemporary mathematical thought, the book's author, Nicolas Bourbaki, was identified as "formerly of the Royal Poldavian Academy," and currently on the faculty of the "University of Nancago." The man's works caught on immediately. First published in France, they were soon being translated into other languages and published throughout the world, making the author quite famous at least among theoretical mathematicians. For many years, however, little was known about the author. The "Royal Poldavian Academy" turned out to be nonexistent, as did the "University of Nancago." Which is not at all surprising, in view of the fact that the same is true of Nicolas Bourbaki.

"Nicolas Bourbaki" is the collective pseudonym for a small group of French mathematicians who, in the mid-1930s, wanted to place their discipline upon new axiomatic foundations, thereby giving it an unprecedented clarity of form and structure. They were teachers, these men, and they wanted math to be understandable, to be teachable. Leader of the project was a diminutive Frenchman by the name of André Weil. He and his co-conspirators Jean Dieudonné, Jean Delsarte, Henri Cartan, and some others decided to get together and write a series of books which they'd publish under a pseudonym.

"So far as I can remember," says André Weil today at the Institute, "the first idea of Bourbaki came—it must have been in 1934—when Cartan and I were both teaching a course in analysis in Strasbourg, and we were constantly discussing 'How should one teach this?,' 'How should one teach that?' And my recollection is that one fine day I told Cartan, 'Look here, we are about half a dozen of us from the École Normale who are teaching those matters in various universities—me in Strasbourg, Delsarte in Nancy, Dieudonné I think in Rennes, or wherever. Let's assemble and settle those matters once for all. Then we won't have to think about it anymore.'"

They'd construct an entire identity for the author, a history, a personality. They'd even concoct an impressive imprint under which his works would be brought out: *Publications de l'Institut Mathématique de l'Université de Nancago*.

"We sent some letters out," Weil continues, "and then we met twice a month in a decent restaurant in Paris—I forget the name but I know very well where it is, it's on the Boulevard St. Michel. Good food but not too expensive, because we were all rather junior people at the time—and so we discussed those matters twice a month, then decided to assemble somewhere during the summer, for a week or two weeks, I forget. Which we did, and that was the first Bourbaki congress in history."

In pursuing their Bourbaki hoax, these young mathematicians were only carrying on the great traditions of their school, the École Normale Supérieure. "There was a tradition of practical jokes, going back to the early '90s of the seventeenth century," Weil says. "For example there was one about Painlevé, the Sorbonne professor. Later he became a prominent man in French politics, and in fact during the war in—1916 or so—he even made it to Prime Minister. But anyway at that time he was a young mathematician, very brilliant, very well thought of, and already—which was quite unusual—he was a professor at the Sorbonne. He was also one of the examiners for admission to the École Normale."

The École Normale, being quite exclusive, required an oral examination for admission.

"So the candidates for the test are standing around there in the corridors," Weil continues, "and a few École Normale students are also hanging around, and one of them—who has been around for a few years—starts talking to one of the candidates. They get to be rather friendly, and after a while the student says to the candidate: 'Well, you seem to be a nice guy, I better warn you about something. You know there's a tradition here in the École Normale about practical jokes'—in French we call them *canular*, that was the École Normale slang for this—so he says 'there's this tradition here, and one thing that has been done several times is that a student at the École Normale poses as an examiner in the entrance examination, and he makes fun of the candidate.'

"So the candidate goes in to take his exam," Weil says, "and he looks at Painlevé and sees before him a very young man, one who almost could be a student, and the candidate says in French: 'You can't put this over on me!' And Painlevé says, 'What do you mean? What are you talking about?' So the candidate says, 'Oh I know the whole story, I understand the joke perfectly, you're an impostor.' And Painlevé says 'I'm Professor Painlevé, I'm the examiner,' and so on . . ."

—Weil is breaking up during the telling of this, laughing and screaming and slapping his knee—

". . . and finally Painlevé has to go and *ask the director of the École Normale to come in and vouch for him!* Aaaaaiiiiii!"

And this is only one of the more *harmless* stories! There was far worse going on back there at the École Normale Supérieure. So it's no surprise, then, that when the cream of the École's young mathematicians got together to write their deadly serious textbooks, their grave axiomatic tomes, they wouldn't simply list their own names as joint authors—"We didn't like that idea," Weil says. So they called themselves "Bourbaki," after an obscure French general, Charles Denis Sauter Bourbaki.

The general, so the story goes, was once offered the chance to be King of Greece but for some unknown reason declined the honor. Later, after an embarrassing retreat in the Franco-Prussian War, Bourbaki tried to shoot himself in the head, but missed. He was the perfect buffoon, and so the group decided to name themselves after him.

After he came to America, Weil taught for a time at the University of Chicago, while some of the other founding members of Bourbaki—Dieudonné, Delsarte, and Claude Chevalley—went to the University of Nancy, in France. The names Nancy and Chicago were combined, and so Monsieur Bourbaki would be at the University of "Nancago."

Bourbaki evidently met a need, for the polycephalic gentleman's books made a profit—and do so to this day—enough to pay for the travel expenses, lunches, and dinners to which the *Bourbachiques* have always treated themselves. The group still exists today, although the membership (there are about ten of them) is different on account of a strict rule to retire at the age of fifty. They work today as they always have. One of the authors writes a book-length manuscript and reads it before his colleagues, all of them well-soused by fine wines, at a joint meeting in Paris. "Certain foreigners, invited as spectators to Bourbaki meetings, always come out with the impression that it is a gathering of madmen," says Jean Dieudonné. "They could not imagine how these people, shouting—sometimes three or four of them at the same time—about mathematics, could ever come up with something intelligent."

But they do. The heavily criticized author goes away to rewrite his manuscript—sometimes it is rewritten by another member—and the process is repeated again and again until the group unanimously declares it ready for the printer. Sometimes a text may undergo seven or eight revisions before publication.

André Weil, although long since retired from Bourbaki, likes to keep on with the *canular*. A recent publication of the Institute for Advanced Study lists the affiliations, honors, prizes, and publications of its august membership from 1930 to 1980. The entry for André Weil reads, in part: "*Honors*: Member, Poldavian Academy of Science and Letters. *Memberships*: Nancago Mathematical Association."

There's yet possible a finer level of magnification than the one we're now looking at, John Milnor explains. In fact, there's no inherent stopping point, and that's part of the mystery of the form in front of us: the more you magnify any portion of it, the more structures you find revealed inside it. It's like an unending series of Chinese boxes.

Milnor enlarges the form once more.

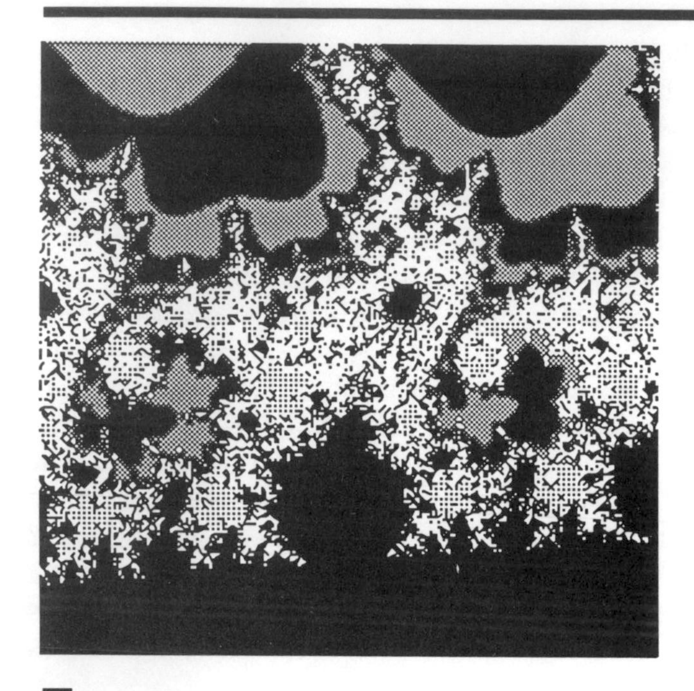

The focal point of the Institute's School of Mathematics is the seminar room. On the first floor of Fuld Hall, just down the hallway from Einstein's office, it's the only true "classroom" at the Institute. It's got fifty or so seats with those little swing-up desks for taking notes, and it has a blackboard at the front. Mathematics seminars are given here at least once a day—often enough twice. Einstein has lectured here, and Gödel, as well as Atle Selberg and André Weil. In the academic year 1984/85, Weil, now professor emeritus, gave a whole course of lectures entitled "Topics in Algebra: Euclid to Bombelli." The lectures are attended by the senior math faculty as well as by the temporary members.

In the audience today, listening to a talk on transcendental number theory, is mathematician Robert Tubbs. In his early thirties, Tubbs has a reddish-brown beard that makes him look like a Russian general or the captain of a whaling ship. He's on leave from his regular position at the University of Texas at Austin, where he specializes in the transcendentals, which for some reason are the object of a new wave of interest among theoretical mathematicians.

"Sounds really glamorous, doesn't it?" Tubbs says after the seminar. "*Transcendental numbers*, wow! Actually it's not so glamorous. All numbers are either algebraic or transcendental. Transcendental means only that a number isn't the root of any polynomial."

Algebraic numbers are those which are solutions to algebraic equations. More precisely, algebraic numbers are solutions of polynomials having rational numbers as coefficients. Thus, the equation

$$x^2 = 1$$

has two solutions,

$$x = 1$$

and

$$x = -1,$$

and so both 1 and -1 are algebraic numbers. To be algebraic doesn't mean to be an integer, since even irrational numbers—many of them—are algebraic. For example, $\sqrt{2}$ is algebraic in spite of the fact that it's irrational, meaning that it goes on forever: 1.4142135. . . . The reason why $\sqrt{2}$ is algebraic is simply that it is the solution of an algebraic equation such as

$$x^2 - 2 = 0.$$

"If you can come up with some equation over the integers that the number is a solution to," Tubbs says, "then the number is not a transcendental."

Since transcendental numbers are defined negatively, as those which aren't solutions to polynomials, the question arises how a mathematician could possibly be sure that any given number *is* transcendental. Couldn't it be that he just hasn't come up with the right equation yet? The answer is that numbers are shown to be transcendental through the use of proofs by contradiction, like Euclid's proof that there is no largest prime.

"You usually assume that there *is* some equation," Tubbs says, "and then you derive a contradiction from that."

Such proofs aren't exactly easy. Johann Lambert proved in 1767 that π is irrational, but it wasn't until more than a hundred years later, in 1882, that Ferdinand Lindemann proved π to be transcendental.

Part of the fascination of transcendental numbers comes from the fact that, whereas almost all numbers are in fact transcendental, only a very few have ever been *proven* to be.

"If you have all the numbers in a bucket, say, all the numbers in the real number line—the rationals, irrationals, the integers, and so on—and you pull one out of the bucket, the probability is one—well, *almost* one—that it will be transcendental. But if you actually want to write a transcendental number down, there's only π and e and a few others."

What is it like, I wonder, to be a pure mathematician, to spend all that time up there among the Forms, among the rationals and the irrationals, the primes and the transcendentals. Why in the world would anyone possibly *care* about these things? So I put the question to Rob Tubbs:

Q: Why do you care about the transcendental numbers?

Tubbs: Well, because they're interesting, and the problems are simple to state. Part of the charm is that the problems that are easiest to state are often the hardest to answer. For example, is $e + \pi$ transcendental? It's *seemingly* impossible to answer, at least for me, right now. Of course if you could *prove* that it was impossible to answer, then that would be a major breakthrough.

Q: Is there really something that's impossible to prove in mathematics?

Tubbs: There might be. There might be things that are independent of the axioms that you're using. There might be very simple little things that are impossible to prove.

Q: Is any of this transcendental number business related to the real world?

Tubbs: Well, it may be related to the real world, at least genetically. I mean all the problems came from someplace originally. The original questions came from geometry, about π for example. So if you were pushed, you could trace this back to geometrical questions. But right now much of mathematics does seem on the surface to be formal symbol-manipulating, proving results about objects that no one has any intrinsic interest in. But in any field there is this pressure on you to publish papers, and—you know—if you happen to get some results, even if they say something strange about some objects that may be of no obvious interest, you want to publish it someplace.

Q: What specific problem are you working on?

Tubbs: The big problem in my field has to do with e, the base of the natural logarithms. Well, it turns out that this function, $f(z) = e^z$

is also an important function, it has great properties. If you plug in for z something like 1, you get e, which is transcendental. And if you plug in 2 you get e^2, which is transcendental. Well, this is the kind of thing that leads a mathematician to ask a general question. For what alphas that I plug in will I always get a transcendental number, using the function $f(\alpha) = e^{\alpha}$? It turns out that you can't answer that completely. I mean one would like to get an exact classification of all the alphas that do this. A partial solution is that if α is algebraic, and not equal to zero, then e^{α} is transcendental.

Q: Are there any alphas that you know *do not* give a transcendental?

Tubbs: If you raise e to the natural log base 2, you get 2, and 2 is not transcendental. So if you take the natural logarithm of anything like 2 or one half, then when you raise e to that power you will get that same number back.

Q: Does this problem go by any particular name?

Tubbs: It's connected to something rather general called Schanuel's Conjecture.

Q: Would you say it's your goal in life to prove Schanuel's Conjecture?

Tubbs: Well, it would be nice to prove Schanuel's Conjecture, but I'd sound like a raving lunatic if I said it was my goal as a mathematician.

Q: Why? It wouldn't be as difficult as proving Fermat's last theorem, or something like that, would it?

Tubbs: It might be, although it's not as glamorous. I mean we know how a proof of this should go. You bring in all these tools from complex analysis and algebraic geometry, but it so happens that these tools are not as good as they should be. But even when we make those tools as strong as we can possibly make them, even then we can't prove this.

Q: But I take it that proving Schanuel's Conjecture would be an important result. Why?

Tubbs: The reason why proving this would be important is that in order to prove Schanuel's Conjecture you would in the process have to prove something else, something bigger, and that would be really great. I mean if you prove it by being tricky, without establishing anything new, then you wouldn't get any recognition, and you probably shouldn't. The assumption is that if you prove one of these great big things like Fermat's last theorem, you'd have to develop a whole new branch of mathematics. But if you prove it by some quirky little two-line proof that everybody had always missed, then people would pat you on the back, and they'd say "God, you're clever!", and so on, but it really wouldn't be as much of an accomplishment.

Q: Do you think you'll ever prove Schanuel's Conjecture?

Tubbs: I don't think I can. I just don't see a way to do it. I don't see a scheme, even. I don't see what the new idea is.

Q: Aside from whether you could prove it, do you think the conjecture is true or false?

Tubbs: Oh, I think it's true. If you assume it, it implies much of what is already known. A lot of the known results follow from assuming that Schanuel's Conjecture is true, and so that's evidence for its being true.

Q: Why do you *want* to prove it?

Tubbs: Well, it would be satisfying to see a proof—maybe only because it would be nice to see what new ideas it would take. What's important in mathematics are the ideas. And in fact it really wouldn't be satisfying to you as the author if you proved it in some little high-school way that no one had ever observed before.

Q: When you're working on a proof and finally establish something, do you think you're discovering something new about the world of mathematical objects?

Tubbs: Oh yes, especially now. I mean writing your Ph.D. thesis you just do it because it's something you can do—it's pretty technical. But once you have the time to think, the free time—which is why the Institute is so great, because it gives you the time without the pressure—you can play with ideas and see how they fit together. At points you come across things that are really nice, and no one else has ever seen them, and that really is *very* satisfying. In fact I think that's why working on Schanuel's Conjecture might be fun.

Q: Are you ever surprised by the results you get?

Tubbs: In my field what we can prove is so much weaker than what we think should be true that we're rarely surprised. Almost every result is a partial solution. Suppose you want to prove that some number is transcendental. Well, there will be some theorem that says "Out of the following twelve numbers, one of them is transcendental," but it doesn't tell you which one. So frequently you just miss the mark. And given that this field is so easy to make conjectures in, and everyone thinks they know what *should* be true, our results are always a little weaker than you can prove. What this shows is that the field needs some new ideas.

Q: Do mathematical objects exist out there, or are they merely human creations?

Tubbs: Well, the integers probably pre-exist in some sense. And the fractions too. . . . But basically I think mathematics is mostly a human creation, so I guess I'm sort of a semi-Platonist. I think that any other creature that's thinking about these problems would come to the same conclusions that we do, and they'd have the same objects in front them. I think that once you have the integers, pure thought would always lead to the rest of mathematics. It's like falling off a log.

Q: If mathematics is a human creation, then why is it so tough to get results?

Tubbs: It's easier to speculate than to substantiate, that's true in any realm of discourse. You can speculate about God and metaphysics and so on, but what can you really prove? Your question may itself be an argument that mathematics is in some sense outside the human thought processes, but I don't think it's there waiting to be discovered. What I think is that it's part of the underlying structure of thought. One reason it's so difficult to prove things is because we mathematicians bought Hilbert's claim that proofs have to be rigorous. If we didn't believe that, we could argue pretty easily and come up with lots of great results.

John Milnor started making mathematical waves as a freshman at Princeton University where he fell under the influence of the mathematician, Albert Tucker. As a freshman, Milnor took a course with Tucker in differential geometry, and he worked on a problem posed by the Polish mathematician, Karol Borsuk, concerning the topology of knots. The question was, how much total bending must a curve undergo in order to form a knot?

"The conjecture, which I was able to prove," Milnor says, "was that if you have a closed curve which forms a knot, then the total amount of bending has to be at least twice 360°, or 720°." Milnor came up with his proof as an eighteen-year-old undergraduate, and a year later he published a full theory of knotted curves. For this he was later awarded a Fields medal.

Milnor has been a professor of mathematics at the Institute since 1970. His office, on the third floor of Fuld Hall, at the front of the building (the worldly side, where from the windows you can actually see some homes, cars, and people in the distance), is an absolute paradigm of disorder and confusion. ("I'm not too very well organized," Milnor says.) His desk, the chairs, the bookshelves, the tables, all these are spilling over with huge piles of loose papers, offprints, mathematical journals, computer printouts, diagrams, pictures, books. You have to be careful where you walk. The only free space is on top of the computers. Milnor doesn't put anything on top of them because that might lead to heat buildup.

There are three computers in Milnor's office, including an IBM PC, a separate terminal hooked up to a mainframe (a VAX 11/780 elsewhere), and another terminal that's not presently in working condition. In an adjoining room there are no less than five more computers, a few Sun Microsystems units, and another IBM or two. One of the Sun machines is connected to an Apple Laserwriter on another table, and cables go crisscrossing over the floor. Here, too, you've got to watch where you step.

Milnor is a completely different type of mathematician from people like Atle Selberg, Rob Tubbs, and André Weil, who don't care about computers in the least. Milnor is an *algorithmic* mathematician, while the others are existence-proof men. This makes all the difference.

Most traditional mathematics is of the classical, existence-proof variety. You begin with definitions and axioms, then you state a theorem and prove it. It's a constant process of theorem and proof, theorem and proof, and the point of it all is to show that some specific mathematical object exists, such as a number having a certain property, or that another type of object doesn't exist, such as a greatest prime number. The paradigm example of existence-proof mathematics is Euclid's *Elements*. To practice this type of math, all you need is pencil and paper, or chalk and blackboard.

For a long time, existence-proof math was the only kind done by mathematicians, but this is the age of computers and all that has changed. There's a whole new world of mathematical objects out there, and they're not the product of axioms and proofs, but of algorithms. At base, an algorithm is defined as a mechanical, step-by-step procedure that will give you a desired result. Algorithms that computers can follow are called programs, and, suitably programmed, computers now routinely give mathematicians the numbers that before they could only dream about. Euclid's existence-proof established that the prime numbers are endless; algorithmic mathematicians, using the computer, have produced the first 50 million prime numbers. An existence-proof told us that π goes on forever; the computer has given us its exact digits—to many millions of decimal places. These number-crunching operations aren't of any special mathematical significance: it's not as if anyone *needs* six million digits of π. But there's one way in which the computer gives mathematicians something that they never had before, and what it gives them is something genuinely new. It gives them *pictures*, direct and immediate visions of mathematical objects.

"If I can give an abstract proof of something, I'm reasonably happy," John Milnor says. "But if I can get a concrete, computational proof and actually produce numbers I'm much happier. I'm rather an addict of doing things on the computer, because that gives you an explicit criterion of what's going on. I happen to have a visual way of thinking, and I'm happy if I can see a picture of what I'm working with."

The picture that Milnor has been showing me all this time—that frenzied electrocuted shape on the screen—is a geometrical figure called the *Mandelbrot set*. Benoit Mandelbrot, its creator, not only invented "the set," as he puts it, "of which I have the honor of bearing the name," but he also, and singlehandedly, invented the entire new branch of mathematics, a field called *fractals*. Some people at the Institute think that fractals are the wave of the future.

Mandelbrot spent a year at the Institute back when he was a young post-doctoral student, in 1953-54. "I had been a post-doc briefly at M.I.T., and I wanted to go somewhere else," Mandelbrot says. "At the Institute, John von Neumann was interested in my work and he invited me to become a member." Von Neumann was spending a lot of his time in Washington as Atomic Energy Commissioner, so Mandelbrot didn't see much of him. He saw more of Oppenheimer, who was then the Institute's director. Oppenheimer once asked Mandelbrot to give a lecture on the work he was doing at the Institute, work that grew out of his doctoral thesis on the theory of games and communication theory. He gave the talk in the mathematics seminar room.

"When I arrived to give the lecture Oppenheimer was already there and so was von Neumann," Mandelbrot says. "And I remember very distinctly a loss of nerve on my part which led to my giving an absolutely abominable lecture, which nobody understood. It was pitifully bad. After it was over someone came up and told me it was the worst lecture he had ever heard! But the day was saved by a very marvelous summary by Oppenheimer and by von Neumann. Oppenheimer and von Neumann regave my lecture, each of them in turn, but much better than I could do it, and so finally it was a very triumphal event, and it turned out very much to my advantage. But it was an occasion that people who were there would remember very well because of my incompetence as a very young man to give a talk to a mixed audience, and then the arrival of my two guardian angels."

Mandelbrot would often go across the campus to see von Neumann's computer, the one that he designed and built on the Institute grounds. But Mandelbrot didn't know exactly what to make of the machine, and so he didn't get involved with it. All that came later, when he was discovering the universe of fractals.

A fractal is a geometrical object whose contours are not smooth, like the lines, curves, and surfaces of classical euclidean geometry, but rather irregular, broken, and jumpy, both on the large and small scales. "I coined *fractal* from the Latin adjective *fractus*," Mandelbrot says. "The corresponding Latin verb *frangere* means 'to break': to create irregular fragments." These irregular fractal shapes are important because, contrary to what we've always been led to believe, the universe is not basically euclidean.

Standard geometry, that is to say, euclidean geometry, really doesn't apply to undoctored nature. As Mandelbrot puts it in his unforgettable manifesto, "Clouds are not spheres, mountains are not cones, coastlines are not circles, and bark is not smooth, nor does lightning travel in a straight line. More generally, I claim that many patterns of Nature are so irregular and fragmented, that, compared with *Euclid*—standard geometry—Nature exhibits not simply a higher degree but an altogether different level of complexity."

The irony of it all is that classical geometry, which historically was developed as a means for measuring large spaces on earth (as is reflected in the word's etymology: *ge*, earth + *metron*, measure), is absolutely no good for measuring earth's natural, undoctored features. "How long is the coastline of Britain?" Mandelbrot asked himself early on in his career. His answer was that there's really no single right answer. Everything depends upon your scale, and upon your standard of measurement. If you took a map and measured the distance between the northern and southern tips of Britain, you'd get a crude estimate of the coastline's length. But if you were to *walk* the coast between the same two points, your answer would be something else again, because you'd be traversing the shore of every inlet and bay, every last cove and peninsula. If you were Mickey Mouse making the same trip, you'd trace out still finer, more irregular pathways, and accordingly travel a much greater distance. If you were an ant, a microbe, . . . the mind boggles.

This is not only of theoretical interest, either. "The lengths of the common frontiers between Spain and Portugal, or Belgium and Netherlands, as reported in those neighbors' encyclopedias, differ by 20 percent," Mandelbrot says. "One should not be surprised that a small country (Portugal) measures its borders more accurately than its big neighbor."

The difference between classical and fractal geometry lies in their opposed notions of dimension. In standard geometry, dimensions come only in whole numbers: a straight line has the dimension 1, a plane surface 2, a solid 3. But fractals, just as they have fragmented, broken edges, also have *fractional dimensions*, strange twilight-zone dimensions such as 1.67, 2.60, and $\log_2(E+1)$.

One might think that dimensions have to be whole numbers: a line is a line and a surface is a surface. But consider a Hilbert curve, which results from progressively dividing a square into smaller squares and connecting their centers with a continuous line. After a few reiterations, the line formed by this process approaches a two-dimensional surface, even though, because the line does not close back upon itself, it is not a true bounded plane (see Figure 4).

To classical mathematicians, shapes such as the Hilbert curve were rather loathsome entities. They were termed "pathological" shapes and deemed not worth studying. But Mandelbrot went ahead and investigated them anyway. Because creating true fractals requires many iterations of complex-number functions, he could produce such forms efficiently only with computers. It was during his career as an IBM research fellow that Mandelbrot got the opportunity to produce a whole universe of fractal shapes. He began by producing artificial coastlines. All he needed was the right formula, which is to say, the right algorithm.

"I thought up a suitable equation," Mandelbrot says, "and in 1973 we rigged up a very clumsy plotter to produce artificial coastlines. . . . Sometimes we had to sit up all night with plotters. But when the first coastline finally came out, we were all amazed. It looked just like New Zealand! Here was an elongated island, there a squarish one, and, off to one side, two specks resembling Bounty Island. . . . Seeing them had

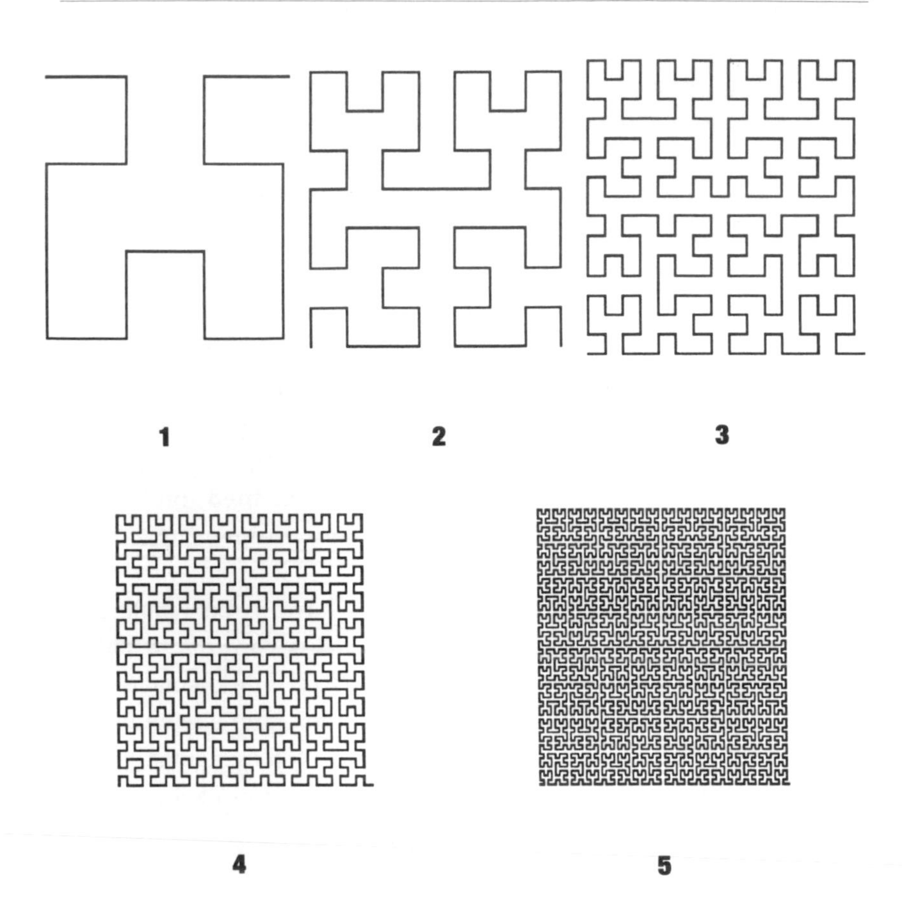

Figure 4 A curve becomes a plane

an electrifying effect on everyone. . . . Now, after seeing the coastline pictures, everyone agreed with me that fractals were part of the stuff of nature."

The beauty of many fractal shapes comes from their property of being self-similar. In other words, the entire shape of a given fractal will be mirrored in each smaller portion. This is the phenomenon that Mandelbrot refers to as "scaling," meaning that the form is the same no matter the scale at which you view the object, as happens in the case of the fractal tree (shown on page 91).

To look at a fractal is to see a visual representation of a simple numerical function that has been reiterated—repeated—again and again. For all its apparent (and real) complexity, the Mandelbrot set is the result of plotting what happens when the utterly simple function ($z^2 + c$) is reiterated over and over again, taking complex (imaginary) numbers as initial values.

"The Mandelbrot set," Mandelbrot himself explains, "is a set of complex numbers which have the property that you take a certain operation and take the square. You take a number z, you take the square of z, and you add c. Then, you square the result, and you add c. You square the result, and you add c, and you do it many times. In principle, every time you get a result you check to see whether you have gone outside a circle of radius 2, and you plot this on a graph. As you keep going, the set becomes drawn with greater and greater detail. But all you're really doing is multiplying something by itself, and adding itself: z squared plus c; everything squared plus c; everything squared plus c."

You might think that you'd get a nice straight line, or a curve, or a spiral. But what you get is this:

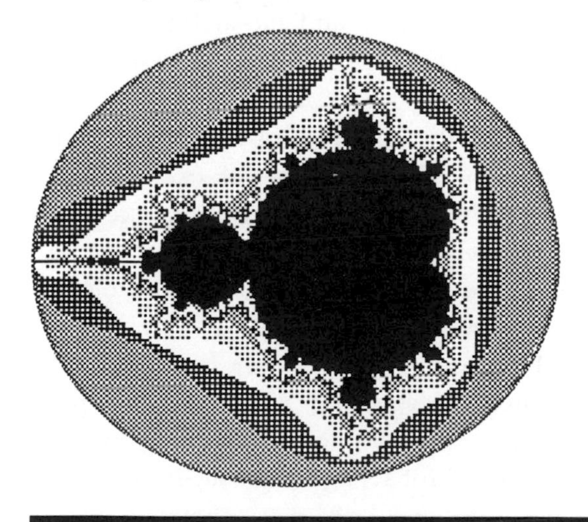

"The Mandelbrot set exhibits, in a particularly extreme fashion," says its inventor, "the phenomenon which is characteristic of fractals, which is that there is a very simple formula, and the result is of extraordinary complication. It's astonishing to find that one line of algorithm, which does not seem particulatrly interesting itself, would lead to something with so extraordinary a structure."

At the Institute for Advanced Study, John Milnor investigates fractals primarily because of their beauty. "I look at these things because they're beautiful in their own right," Milnor says. "For some people the main motivation is that study of these things is likely to be useful. For me personally, usefulness is just a happy by-product." Milnor nevertheless acknowledges that the study of fractals has applications.

"Fractals may be tied up with different real-world problems, such as understanding the nature of dynamical systems, finding out what turbulence is, for example," Milnor says. "Also fractals may give us a better model of the human lung system than conventional geometry does. Think of the very fine blood vessels and air channels interconnecting with each other in a complicated pattern. This doesn't make any sense at all from the point of view of classical geometry, where you study smooth, differentiable objects, but the lung's structure can be described very fruitfully as a type of fractal set."

Mandelbrot sees fractals everywhere in nature, and has produced equations that, when graphed by the computer, mimic the fractal structure of natural phenomena as diverse as trees, rivers, the human vascular system, cloud billows, the eye of Jupiter, the flow of the Gulf Stream, mammalian brain folds. Taking man-made shapes into account, Mandelbrot shows that the folds of theatre curtains and the architecture of the Eiffel Tower have their counterparts in abstract fractal geometry. Fractals, he says, "are the very substance of our flesh!"

With fractals explaining the structure of natural phenomena from the level of fleas to galaxies, the question arises as to the actual scale, in nature, of fractal geometry. John Milnor says: "I think it certainly breaks down when you get to the level of molecules. There you need a different kind of geometry. But if you're on a scale of up to ten molecules or so, fractal geometry might apply from there on upward."

At the moment Milnor can only make guesses about the answers to some of the problems he's investigating. "The particular problem I'm working on right now," he says, pointing to the screen, "is whether, when you blow up a part of the Mandelbrot set again and again, you come up to a kind of limit. So far the picture only gets more and more complicated . . . and there's a great deal one doesn't understand about that."

Benoit Mandelbrot named one of his more symmetrical and pleasing fractal shapes the *San Marco dragon*. "This is a mathematician's wild extrapolation of the skyline of the Basilica in Venice, together with its reflection in a flooded Piazza."

It's relatively easy to program on a computer. You just type in a simple algorithm, a program twenty or so lines long, and you let it run. And there before your eyes comes bubbling up out of the cathode-ray depths a disembodied, ghostly shape. It does look like the cathedral in Venice reflected on the waters, and it's just a little bit mystical, the way it appears, dot by dot. The shape rises up with the electrons, and then floats there on the computer screen. And then you can begin to see what it is that mathematicians find beautiful in their pure disembodied objects of the intellect. Is this what the Forms look like . . . up there in the Platonic Heaven?

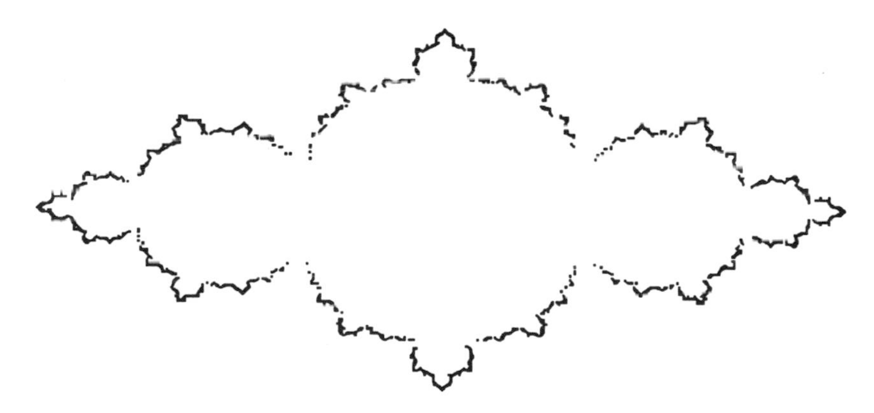

Heretics

Chapter 5

Good Time Johnny

The roulette wheel spins, the white acetate ball goes one way, the wheel itself the other, but all eyes are on the ball as the numbers—alternating black and red squares—rush by in a blur. The room is totally quiet, the only sounds the whir of the air rushing past the wheel frets and the murmur of the ball as it careens its way around the rotor like a moon around a planet. Like many gamblers, the watchers think they know which number's going to come up—or, in this case, at least approximately which one. Roulette wheel rotors are divided up into eight sections, or octants, and the watchers are betting that the ball will land somewhere in the fifth octant, which covers the numbers 18, 31, 19, 8, and 12.

The rotor slows and the ball starts down toward the metal diamonds on the sides of the wheel. If the ball hits one of them, its trajectory will be altered, randomized somewhat, making the fifth octant not such a good bet after all, but the ball drops past the diamonds without touching any of them, and settles in toward the slots below. It arcs over the face of the numbers—alternating even and odd, black and white—and as it does so the fifth octant comes around right on cue, the ball drops into the cups—*click, click, click,* it bounces in and out—and then comes to rest exactly where it's supposed to, at number 19. It's cradled there in its cup like an egg in an egg carton as the wheel slowly glides to a stop.

The watchers are happy enough, but they don't stomp and cheer as amateur gamblers might, nor do they collect any money, for, although it's a regulation roulette wheel that they've got in front of them, made by B.C. Wills, of Detroit, and purchased from Paul's Gaming Devices in Reno, Nevada, this is not Reno, Las Vegas, or even Atlantic City. This spin of fortune took place in Princeton, New Jersey, at the Institute for Advanced Study. In a hidden aerie up on the third floor of Fuld Hall, two floors above the very spot where Albert Einstein used to sit and think . . . about how God doesn't play dice with the universe, two young physicists are perfecting a new gambling system.

Standing on either side of the wheel are J. Doyne Farmer, currently an Oppenheimer fellow at the Los Alamos National Laboratory, and Norman Packard, Farmer's childhood friend from back in Silver City, New Mexico, where they both grew up. Packard is now a long-term member of the Institute for Advanced Study, one of a small number of scientists who make up Stephen Wolfram's complex systems group. Farmer, trained as an astrophysicist, and accustomed to thinking about tiny spheres orbiting large spinning bodies, had written a computer program that he hoped would simulate the dynamics of a roulette ball so closely that he'd be able to predict its exact landing spot. The idea was to take the program, inside a concealed computer, into the Las Vegas casinos and make a fortune. Packard and Farmer had tried it out at Vegas several times, and they seemed to be averaging a 40 percent margin over the casino, but then a series of technical glitches, having to do with the computer hardware, not with the program, forced them out of action. Later, word of the project got out in the popular press—in the book *The Eudaemonic Pie,* portions of which had been serialized in *Science Digest* in 1985—and that pretty much killed off their chances of making a small fortune, at least for a while. But the two of them are back at it again now, although this time it's as much from a sense of closure—a sense that they have to finish the project for its own sake—as it is to go out and break the banks at Monte Carlo.

The amazing thing about this roulette wheel, though, isn't that Farmer and Packard can predict outcome of plays much better than chance, but that the experiment is taking place in the hallowed precincts of Fuld Hall . . . and nobody is raising a ruckus. The One True Platonic Heaven has been turned into a gambling casino laboratory, and these guys are getting away with it! The roulette wheel, true enough, isn't out in the open where just anyone might happen onto it during a day's cerebral work at the Paradise for Scholars. But it's not as if Packard and Farmer have made any overt attempt to hide it from prying eyes, it's just that outsiders never come all the way up here to the third floor of Fuld Hall. And if they did, all they'd see is a closed door which, if they opened it, would reveal a roomful of computers and other doorways going off into other rooms, like a maze. The roulette wheel is in one of these back rooms, which is just as well, because if the Institute regulars—especially those in social science or in historical studies—if they ever discovered that a couple of scientists were up there . . . *playing roulette!* . . . well, there'd be hell to pay. Just think of what happened the only other time anyone at the Institute conducted an experiment.

The guilty party on that occasion was John von Neumann. Here, in this most celestial of all ivory-tower environments, where the heaviest piece of equipment is a piece of chalk, where the loudest noise is that of a few papers rustling in the library, von Neumann went ahead and constructed a new species of electronic computer. No imaginary abstraction, this was the real thing, a nuts-and-bolts, angle-iron-and-sheet-metal *machine*. It had a stack at the top, a flue, an exhaust pipe up which the heat of all the glowing filaments and vacuum tubes inside could escape. Just like a steam engine.

To the Institute regulars this stuff was unthinkable. These Monster Minds had come to Princeton specifically to get away from the crass world of noise and machines, to a place where they could think their deep thoughts in peace and quiet . . . and here was Johnny von Neumann turning their unworldly paradise into . . . *a shop!* Using their monastic Institute facilities to build . . . *an appliance!*

This was no way to behave at the Institute, the One True Platonic Heaven. It was unworthy. It was heretical. It would have to be stopped, and ultimately the Institute regulars got rid of the thing. But that was after von Neumann had died. Although they hated and had no use for his ugly electronic contraption, nobody could stay mad at von Neumann for very long. He was too likeable. He gave these immense parties, the best ones in Princeton. He loved women and fast cars. He loved jokes, limericks, and off-color stories. He loved noise, Mexican food, fine wines, and money. You just couldn't hate a man like that, and so the Institute regulars made allowances and exceptions for von Neumann that they would not have considered for anyone else. For all his dirty-handed messing with computers, he was still one of the high-minded luminaries, one of the immortals, one of the gods that trod upon Earth. "The story used to be told about him in Princeton," Herman Goldstine has written, "that while he was indeed a demi-god, he had made a detailed study of humans and could imitate them perfectly."

Indeed, von Neumann's work on computers and cellular automata wasn't even half of his life's work; it was more like a fifth, or even less. He had talent for creating whole new branches of mathematics, like game theory, for example. To von Neumann, proving the ergodic theorem was not inherently a more worthy activity than predicting the weather, building a computer, or teaching the titans of commerce how to take advantage of game theory the better to position themselves in the dog-eat-dog business world. At Los Alamos during the Manhattan Project, Enrico Fermi used to taunt Edward Teller: "Edward-a how come-a the Hungarians have not-a invented anything?" But von Neumann, who was Hungarian himself, helped invent the implosion mechanism for the first atom bomb and then, along with Teller, Stanislaw Ulam, and others, went on to invent the H-bomb. It wasn't quite *right*—indeed it was horrible, truly dismaying!—to see this Institute professor building computers and making bombs as happily as he invented mathematical disciplines and raked in money from his various consulting jobs. But who could hold it against Johnny? Nobody. He was just too much of a good-time boy.

It was a portentous year for science and technology. In February of 1903, the *New York Times* carried its first story on "the much talked of radium," and news of radioactive elements was broadcast around the world. Later, in October, the *St. Louis Post Dispatch* ran a story which said of radium that, "Its power will be inconceivable. By means of the metal all the arsenals of the world would be destroyed. It could make war impossible by exhausting all the accumulated explosives in the world . . . It is even possible that an instrument might be invented which at the touch of a key would blow up the whole earth and bring about the end of the world."

Of course a few good things happened that same year. On December 17, 1903, on a beach in North Carolina, the Wright brothers made their first four-powered, controlled flights across the sand, starting us down the road to space. Eleven days later John von Neumann was born in Budapest. He would take us into the age of computers, robots, and artificial intelligence. The son of a well-to-do banker, Johnny was not a late bloomer like Einstein. By the time he was six Johnny was dividing two 8-digit numbers in his head and joking with his father in ancient Greek. Two years later he was doing calculus and showing off his photographic memory by reading a page of the Budapest telephone directory and repeating back the names, addresses and phone numbers with his eyes closed. Once, when his mother was sewing, she paused for a moment and stared off into space. The boy, looking at her, asked, "Mother, what are you calculating?"

Johnny enrolled at the University of Budapest, which he used mainly as a base of operations and refueling point for his travels—to Berlin, to hear Einstein lecture on statistical mechanics, to Zurich where he enrolled in the chemical engineering program at the famed ETH, and of course to Göttingen, where he would study with the renowned mathematician David

Hilbert. At the age of twenty-two von Neumann crowned all this feverish activity by getting two degrees, an undergraduate diploma from ETH in chemical engineering, and a Ph.D. from the University of Budapest, *summa cum laude* in mathematics—with minors in experimental physics and chemistry thrown in for good measure.

When he arrived at Göttingen, quantum mechanics was just coming onto center stage. The challenge was to give a consistent mathematical description of the atom, one that would encompass the rival theories advanced by Werner Heisenberg and Erwin Schrödinger. After getting his degrees, von Neumann applied himself to the problem of combining the two, and during the years 1925 to 1929 wrote a series of papers that he would in 1932 publish as his first book, *Mathematical Foundations of Quantum Mechanics*. Today, more than fifty years later, it is still in print.

The key to von Neumann's account of the quantum is his use of "Hilbert space," a notion Hilbert had invented for the purpose of studying equations with infinitely many variables. If you wanted to solve a pair of simultaneous equations, such as

$$x - y = 1$$

and

$$x + y = 7,$$

then you could find the values of x and y in either of two ways. You could use elementary algebra and find a solution arithmetically, or you could use the techniques of analytic geometry. To do this you'd plot both formulas on the same pair of axes; if there was a valid solution to the two equations simultaneously, then the resulting curves would intersect at a common point, whose x and y coordinates would be the correct values of the two unknowns.

The same process can be applied to formulas having more unknowns, so that if you had the equation

$$x^2 + y^2 + z^2 = 1,$$

then adding a z-axis will create a three-dimensional space, onto which this new equation can be plotted. Doing so gives in this case a sphere of radius 1, with its center at the common intersection of the three axes.

An equation can be plotted even if it has more than three unknowns, but doing this means leaving the ordinary world of three dimensions and entering into the twilight-zone realm of Hilbert space. More variables mean more axes in the graph: an equation with five unknowns, for example, will describe a five-dimensional sphere, or "hypersphere." What Hilbert did was to extend this progression to cover equations with infinitely many variables, the geometric representation of which would require a space with infinitely many dimensions. Such a space, it's true, is not a physical space—so long as physical space is understood to be composed of the usual three dimensions. Nevertheless, mathematicians and physicists use this infinitely dimensioned Hilbert space routinely, especially in the context of quantum theory, to solve real-world problems. That they do so is largely owed to von Neumann.

In the mid-1920s, two opposed interpretations of quantum phenomena were making things tough for physicists. One was Heisenberg's matrix mechanics; the other was Schrödinger's wave-function theory, and physicists didn't know which of the two to believe. According to Werner Heisenberg, a quantum system's attributes are expressed in matrices—rectangular arrays of numbers such as those on a bingo card or on the periodic table of the elements. Each matrix of numbers represents a different attribute, and so there's one matrix for a quantum's energy-level, another one for position, another for momentum, and so on. Heisenberg used a matrix of numbers—rather than single numbers—because he viewed a particle's attributes as being inherently uncertain and indefinite. A particle is more like a smear in space than like a point on a line, he said, and so its position cannot be represented by discrete integers, but only by whole arrays of them. The different numbers of the array correspond to the different probabilities of a particle having those particular values.

Erwin Schrödinger, on the other hand, maintained that atomic states should be understood as waves of matter. An electron orbiting a nucleus, in his view, would not describe a smooth, circular pathway like a planet going around the sun, but would take a sinusoidal, roller-coaster ride around the atomic core. Other quantum particles would likewise be represented as waves, and their laws of motion would be expressed as wave equations.

At this point John von Neumann entered the fray and joined the two theories together. The key to it all was Hilbert space. Von Neumann showed that if atomic states were understood as vectors (or arrows) in infinitely dimensioned Hilbert space, then the arrows' rotations would correspond equally well to the numerical entries of Heisenberg's matrices and to the wave functions in Schrödinger's theory. Johnny worked all this out in a new axiomatic mathematical framework, one which made the apparently random behavior of quantum particles seem almost logical.

Having done all this by the age of twenty-six, von Neumann acquired an international reputation, and in the fall of 1929 Oswald Veblen, then still of Princeton University's mathematics department, invited Johnny to come to Princeton and give a series of lectures "on some aspect of the quantum theory." Von Neumann accepted, and after a short time in the country he decided that he and the United States were tailor-made for each other. Here was the land of optimism, pragmatism, and can-do. The people were outgoing, friendly, informal; best of all, they liked to have a good time, just like he did. Of course America did lack some of the old-world comforts, such as the cafés, the little bistros where you could sip espresso and smoke cigars and discuss the status of the ergodic theorem for hours on end. For a while von Neumann thought about opening a European-style taverna in Princeton, but he never went through with the idea. He did do the next-best thing, however: he threw his parties. To hear Institute old-timers tell it, they were like small operettas.

"They were unbelievable," says one of von Neumann's old friends. "The stories you read about those parties, they're not exaggerations. Von Neumann was a fantastically witty person, a lusty person, he was fatter than I am. He knew how to have a good time." These evening affairs were held regularly, at least once a week, sometimes twice, at von Neumann's big white clapboard house at 26 Westcott Road, where uniformed servants used to come around with the drinks. There was dancing, smoke, loud laughter, and camaraderie. "Those old geniuses got downright approachable at the von Neumanns'," a friend remembers.

When the Institute for Advanced Study was getting underway at Princeton it was only natural that von Neumann should be invited to become a member, and so along with Einstein, Veblen, and Alexander, Johnny became an Institute professor. He was a youngster in a group of old men. "He was so young," says one Institute member, "that most people who saw him in the halls mistook him for a graduate student."

At an Institute bristling with great intellects, von Neumann had the fastest mind by a large margin. "He had the kind of mind," says Julian Bigelow, who worked with him on the computer project, "that if you go in to see him with an idea, inside of five minutes he's five blocks ahead of you and sees exactly where it's going. His mind was just so fast and so accurate that there was no keeping up with him. There was nobody on earth, as far as I'm concerned, who was in his category."

Lots of mathematicians say that they aren't particularly good with numbers, that they can't add, subtract, multiply, or divide faster than anyone else. But von Neumann was a human adding machine. "When his electronic computer was ready for its first preliminary test," says Paul Halmos, one of Johnny's assistants at the Institute, "someone suggested a relatively simple problem involving powers of 2. (It was something of this kind: what is the smallest power of 2 with the property that its decimal digit fourth from the right is 7? This is a completely trivial problem for a present-day computer: it takes only a fraction of a second of machine time.) The machine and Johnny started at the same time, and Johnny finished first."

You had to be a very quick note-taker indeed if you were going to follow one of von Neumann's lectures. During his seminars (Fuld Hall's seminar room was right across the hallway from his office) he'd write dozens of equations on the blackboard, jamming them all into a two-foot square space off to one side. As soon as he was finished with one formula he'd zip it away with the eraser and replace it with another one. He'd do this again and again, one right after the other—an equation and *zzzip*, another one and *zzzip*—and before you knew it he'd be putting the eraser back on the ledge and brushing the chalk dust from his hands. "Proof by erasure," his listeners called it.

The man also had a true photographic memory, never forgot a thing. "As far as I could tell," says Herman Goldstine, "von Neumann was able on once reading a book or article to quote it back verbatim; moreover, he could do it years later without hesitation. . . . On one occasion I tested his ability by asking him to tell me how the *Tale of Two Cities* started. Whereupon, without any pause, he immediately began to recite the first chapter and continued until asked to stop after about ten or fifteen minutes."

In the best supergenius tradition, von Neumann had his share of eccentricities. For one, he always dressed like a banker, no matter what the circumstances. He and his wife once went on a trip to Arizona, where they visited the Grand Canyon. Being a good-time boy, Johnny of course wanted to go down to the bottom, on one of those mule train rides. Everyone else was dressed *á propos*—short sleeves, chaps, cowboy boots, sombreros, and so on, but not von Neumann. Certainly not. He was up there on his horse or mule or whatever it was, in his standard white shirt and tie, suit jacket, and display handerkerchief. Apparently he believed in suffering for style. And then there was the matter of his developing the proper accent. Despite everything, von Neumann didn't want to be too American. "He pronounced 'integer' as 'integher,'" Herman Goldstine explains—but every now and then he would say it right. "But then [he] quickly corrected himself and again said it in his own style."

And there were the absent-minded professorisms. Von Neumann's wife Klara recalled that once when she was sick, "I sent him to get me a glass of water; he came back after a while wanting to know where the glasses were. We had been in the house only seventeen years." (Well, they had servants. They knew this type of thing.) Another time Johnny drove out of Princeton one morning for an appointment in New York City. But halfway there he forgot who he was supposed to see, and so he phoned his wife wanting to know, "Why am I going to New York?" What a guy!

It was inevitable that the fastest mind in Western civilization should sooner or later meet up with "the electronic brain." The ENIAC—the Electronic Numerical Integrator and Computer—was being built in Philadelphia, just fifty miles down the road from Princeton. The convergence of these two number crunchers was one of those fateful world-historical moments which resulted from a chance encounter between two men. The men were von Neumann and Herman Goldstine, both of whom were doing work for the U.S. Army's Ballistics Research Laboratory at the Aberdeen Proving Grounds, in Maryland. The ENIAC was being built for the Army, for whom it was going to calculate missile trajectories and firing tables like greased lightning. Goldstine used to shuttle back and forth between Aberdeen and Philadelphia, and one day, in August 1944, while he was waiting for the train, who should wander up the station platform but John von Neumann.

"Prior to that time I had never met this great mathematician," Goldstine remembers, "but I knew much about him of course and had heard him lecture on several occasions. It was therefore with considerable temerity that I approached this world-famous figure, introduced myself, and started talking. Fortunately for me von Neumann was a warm, friendly person who did his best to make people feel relaxed in his presence. The conversation soon turned to my work. When it became clear to von Neumann that I was concerned with the development of an electronic computer capable of 333 multiplications per second, the whole atmosphere changed from one of relaxed good humor to one more like the oral examination for the doctor's degree in mathematics."

A few days later von Neumann was in Philadelphia poring over what there was of the ENIAC. "At this period," Goldstine says, "the two accumulator tests were well underway. I recall with amusement Eckert's reaction to the impending visit. [J. Presper Eckert was, together with John Mauchly, the ENIAC's co-inventor.] He said that he could tell whether von Neumann was really a genius by his first question. If this was about the logical structure of the machine, he would believe in von Neumann, otherwise not. Of course, this *was* von Neumann's first query."

Six months after this first meeting between The Mind and The Machine, von Neumann was planning to build his own computer at the Institute for Advanced Study. First on the agenda, though, was to get a clear picture of the ENIAC's drawbacks, of which there was no shortage. For one thing, it was too big. In fact it was worse than big, it was colossal, a veritable dinosaur of tubes and wiring. It was 100 feet long, 10 feet high, and 3 feet deep. It had over 100,000 parts, including 18,000 vacuum tubes, 1,500 relays, 70,000 resistors, 10,000 capacitors, and 6,000 toggle switches. There seemed to be no end to the thing, and von Neumann used to joke that just keeping it going was "like fighting the battle of the Bulge every day." When the machine once ran for five days without a single tube failing, the inventors were in hog heaven.

The ENIAC consumed so much power that, according to legend, every time it was turned on, the lights dimmed all over West Philadelphia. From a functional standpoint, though, the machine's size, failure rate, and power requirements were as nothing compared to the demands imposed by its relatively hard-wired programming. Unlike modern general-purpose computers which can switch from word processing, to graphics, to game-playing at the flick of a floppy disk, the ENIAC had been designed primarily to do one thing, and that was to compute firing and bombing tables. Getting it to do anything else was a major production. Every time you wanted to give the machine a new type of problem, you had to go around resetting switches and replugging cables one at a time, all by hand. Since the machine had thousands of individual switches and hundreds of external cables and plugs, it could take up to two or three days for a couple of technicians to set up the ENIAC to run a problem that would take it only a matter of minutes to compute.

This way lay madness. There was an idea going around that would change the concept of a computer fundamentally, an idea now known as stored programming. Its origins are obscure. Some computer historians say that it came from von Neumann himself, others say Mauchly and Eckert, while still others trace it back to the British mathematician Alan Turing. (Von Neumann had met Turing at Cambridge University in the summer of 1935, and later Turing came to Princeton to get his Ph.D. Von Neumann offered the younger man a post at the Institute as his research assistant, but Turing declined this in favor of returning to Cambridge.) But wherever it came from, von Neumann seemed to be the one who took the idea of stored programming and transformed it into a working system. His idea was to put the machine's programming *inside* the machine, not in the form of internal wiring, but rather in the form of electrical charges and impulses. This would be an advantage because it would allow you to control and alter the machine's operations without repositioning its external wiring, switches, and connections.

The concept of a machine being controlled from the inside, however, ran against conventional wisdom and common sense. Machines had always been controlled from the *outside*, by means of knobs, levers, buttons, and so on. Even machines that were programmed, like the Jacquard loom, were controlled by physical objects—punchcards or tapes—that were outside of, and often physically separate from, the machine itself. To argue—as von Neumann did—that a machine could be controlled from the inside by impalpable electrical impulses, this required a major leap of the intellect.

Von Neumann decided that the basic functions of a computer—addition, subtraction, and so forth—could be hardwired into the machine, made a part of its physical structure. But the order and combinations in which it would perform the functions, these things could be softwired. To get a machine to work different problems, you wouldn't have to run around throwing switches and replugging cables. *You wouldn't have to change the machine*. You'd leave the machine as it was and simply *change its instructions*. "Once these instructions are given to the device," von Neumann said, it would "carry them out completely and without any need for further intelligent human intervention." In goes the problem, out comes your answer. No muss, no fuss.

In the spring of 1946 Johnny got serious about building a computer of his own at the Institute for Advanced Study. There were only two obstacles: money, and the approval of the Institute's faculty. The money was the easy part. The hard part was getting the Institute regulars to let him build a machine on their hallowed grounds. Even in the School of Mathematics itself, the computer project was not what you could call a big hit.

The School called a meeting to discuss the problem and, according to the minutes, "The discussion considered the effect of such activities upon the progress of mathematics and upon the general atmosphere of the Institute. The personal views expressed ranged from that of Professor Siegel, who, in principle, prefers to compute a logarithm which might enter into his work rather than to look it up in a table, through that of Professor Morse who considers the project inevitable but far from optimum, to that of Professor Veblen who simplemindedly welcomes the advances of science regardless of the direction in which they seem to be carrying us." (The minutes were taken by Veblen himself, who was known to slip in an acerbic comment or two.) Einstein didn't seem to care one way or another—a computer, he joked, wouldn't bring him any closer to a unified field theory.

Faculty members in other departments, such as the School of Humanistic Studies, were even less receptive. Even today, some of the school's old-line members are aghast at the very idea of *building something* at the Institute for Advanced Study. Harold Cherniss, a specialist in ancient Greek philosophy, became an Institute professor in 1948, when the machine was already under construction. "When you look back," Cherniss says today, "there are obviously strong arguments in favor of building the machine. But I still would have been against it. The computer had nothing to do with the purpose for which the Institute was founded. The computer was a *practical* venture, but the Institute is not supposed to be practical."

Frank Aydelotte, on the other hand, who had by now taken over the reins from Abraham Flexner, was quite ready to take the Institute off in more practical directions. This has happened again and again over the years. It's as if, in spite of their lofty position at the throne of the Platonic Heaven, the Institute's directors don't really feel deep down in their bones that it's altogether healthy for such a large bunch of people to be doing nothing but sitting around thinking.

Aydelotte, at any rate, told his board of trustees that, no matter how messy-handed it all might be, the computer project was one thing the Institute ought not to pass up. "I think it is soberly true to say," he said at a trustee meeting, "that the existence of such a computer would open up to mathematicians, physicists and other scholars areas of knowledge in the same remarkable way that the two-hundred-inch telescope [at

Mount Palomar, then under construction] promises to bring under observation universes which are at the present moment entirely outside the range of any instrument now existing." The computer would be a physical thing, true enough, but we can build it here anyway because its *justification* is theoretical. "It seems to me," he said, "very important that the first instrument of this quality should be constructed in an institution devoted to pure research."

Well, how could anyone resist? Here was the fastest mind in western civilization, the man to whom neuron and diode were on speaking terms, asking for a mere $100,000 so that he could get on with his work. Von Neumann was already thinking about the relation between mechanical and biological brains, and who could tell what might emerge from this? And of course there was the not inconsiderable fact that the man behind it all would be none other than our very own good-time Johnny.

So they gave him his $100,000. But that wasn't all. The Radio Corporation of America came in with more money, and so did the Army Ordnance Department, and the Office of Naval Research and Development, and the Atomic Energy Commission. Money was no problem; in fact, compared to gaining the support of the Institute faculty, getting the money was kid stuff.

A year and a half after von Neumann and Goldstine met on the Aberdeen railway station, Johnny was hiring staff for the ECP: the Electronic Computer Project of the Institute for Advanced Study. He had already gotten Goldstine to leave the ENIAC project and come to the Institute. Then he got Arthur Burks. Burks was a rare man, a Ph.D. in philosophy who also understood electrical circuitry. But in addition there had to be the people who would actually *build* the thing—with their hands. Von Neumann would supply the grand ideas, the goals, the general design principles, but wielding a solder gun was clearly not part of his repertoire. He needed a chief engineer.

M.I.T. mathematician Norbert Wiener recommended Julian Bigelow. Bigelow had an electrical engineering degree and had worked for a while for IBM, and then came to M.I.T. during the war years to work as Wiener's assistant. Wiener and Bigelow were designing an automatic aiming mechanism for anti-aircraft guns. The core of the thing was a data processor that would collect information on the aircraft's flight path and then make a projection about how to aim the gun. If everything worked correctly, the shell and the plane would arrive at the same place at the same time, producing a splendid *auto-da-fé*.

In January 1946 Bigelow came down to Princeton for an interview with von Neumann. He was a couple of hours late. Bigelow was driving down from Massachusetts in the little 1937 Willys he had at the time, but the car wasn't in the best shape and required frequent on-the-road tweaking and adjustments to keep it going. Finally, just as von Neumann was about to give up hope, a decrepit vehicle pulls up to the front of his house and dies amid several noisy backfires. Julian Bigelow got out of the car and walked up to the house.

"Von Neumann lived in this elegant lodge house on Westcott Road in Princeton," Bigelow says. "As I parked my car and walked in, there was this very large Great Dane dog bouncing around on the front lawn. I knocked on the door and von Neumann, who was a small, quiet, modest kind of a man came to the door and bowed to me and said, 'Bigelow, won't you come in,' and so forth, and this dog brushed between our legs and went into the living room. He proceeded to lie down on the rug in front of everybody, and we had the entire interview—whether I would come, what I knew, what the job was going to be like—and this lasted maybe forty minutes, with the dog wandering all around the house. Towards the end of it, von Neumann asked me if I always traveled with the dog. But of course it wasn't my dog, and it wasn't his either, but von Neumann—being a diplomatic, middle-European type person—he kindly avoided mentioning it until the end."

Von Neumann told Bigelow that he wanted to build an entirely new computer, a very high-speed, truly general-purpose, stored-program machine. "To begin with," Bigelow says, "it would be parallel, it would be stored-program, it would be very simple in that it would have a small number of arithmetic operations—addition and subtraction—which it could do very rapidly. Von Neumann felt that these would do the whole job since you could program multiplication and division out of conditional add-subtracts. The idea was to get the thing running as fast as possible, get it as high-speed as you can, and then the programming will take care of the rest. Von Neumann described the kinds of speeds he wanted to get—bit transfers in a microsecond or so. But in the end he came around to thinking that it might be more efficient to program multiplication in too."

The computer began in the basement of Fuld Hall, in the boiler room, in June 1946, and soon von Neumann, until then a mathematical physicist, was up to his ears in electronics and writing letters that said things like: "There exist two recent (1944) midget pentodes which may be of interest to us: 6AK5 and 6AS6. . . . Both have sharp cutoff on the control grids. 6AK5 has inner connection between supressor and cathode; 6AS6, however, brings the supressor out separately and has a sharp cutoff on the supressor too: −15v for +150v on the screen."

The Institute wanted to put the project into a separate building, to get it out of sight and out of mind of the Institute regulars, but there was a problem about getting permission from the city. After all, this was a residential neighborhood, one of the richest and most exclusive in Princeton, and the good burghers didn't like the idea of a machine shop—"a computer factory"—springing up in their back yards. So there was a town meeting. "And there was this idiot from RCA Laboratories, with a Ph.D. in chemistry, no less, and he got up in that town meeting and said that we don't want the building down there because it would make too much noise. *But you couldn't have heard it,*" Bigelow says. "If you were standing in the street outside *you could not tell* if we were in there building the machine that day or not."

The engineers spent a year in the boiler room, making their test apparatus and designing the prototype. By January 1947 the Institute had gotten the city's permission to put up the new building, and the structure was going up across the campus. It was a large, plain, one-story building, not in the Georgian style at all, and separated from the rest of the Institute by space, design, and atmosphere. The computer crew moved into the ECP building that summer.

The prototype worked the first time out. "It worked so well," Bigelow says, "that when we first turned it on it didn't need any adjustments or trimming." Then they laid out the complete 40-stage unit. "Von Neumann would put half-finished ideas on the blackboard and Goldstine would take them back down and digest them and make them into something for the machine. On the other hand, von Neumann often had only the foggiest ideas about how we should achieve something technically. He would discuss things with me and leave them completely wide open, and I would think them over and come back with an experimental circuit, and then my group would test it out."

Although the I.A.S. machine was a stored-program computer, its programs were not written in any of today's high-level languages, like BASIC or Pascal. They were written directly in machine language, composed of long strings of ones and zeroes. To get the machine to do what could be done on a modern computer today simply by pressing the backspace key required entering a machine-language phrase on the order of 1 1 1 0 1 0 1. "There was no assembly language, even," Bigelow says, "none of the tricks that we now have. This was a case where von Neumann was so clever technically that he had no problem with it. And he couldn't imagine anyone else working with a computer who couldn't program in machine code."

To add insult to injury here in this realm of peaceful contemplation, the machine's shakedown test was not something innocuous like running a program for finding the first 5,000 prime numbers. Oh, no. It would be nothing so harmless and prosaic. Von Neumann was involved in H-bomb work at Los Alamos—many of the offices in the ECP building were put there specifically for the use of visiting Los Alamos scientists—and Johnny had the idea that one of the calculations needed for the thermonuclear reaction should be tried on the I.A.S. computer. The computation required was monumental, the largest ever done up to that time, by man or machine, taking more than a billion elementary arithmetical and logical operations just to find out whether the reaction would propagate as desired. So the first problem was to figure out whether the H-bomb would explode. The answer was yes.

"It was computed in the summer of 1950 by Marshall Rosenbluth," Bigelow says, "while the machine had clip leads on it. We had engineers there to keep it running and it ran for 60 days, day and night, with very few errors. It did a nice job. And it was a very historic computation."

Later, of course, when the Institute formally unveiled its computer, in June of 1952, the demonstration problem was one that would be acceptable to any pure mathematician. It dealt with Kummer's conjecture, a problem in prime number theory. To celebrate the unveiling, von Neumann gave yet another party. There in the von Neumann living room was a scale model of the Institute for Advanced Study computer. It was sculpted in ice.

Von Neumann's machine was fully automatic, digital, and all purpose. It was a stored program computer whose inner architecture became the standard for a later generation of commercial machines. By any practical measure—not the yardstick to be applied at the Institute, of course!—von Neumann's computer project was a walkaway triumph. While the machine was operating it worked on problems in abstract mathematics, in physics, and in numerical meteorology. It did computations on the internal structure of stars, and on the stability of orbits in particle accelerators. It was a true all-purpose machine.

More important than the individual problems it worked, the von Neumann computer was the occasion for a vast outpouring of papers from the Institute, pioneering works on the theory and practice of machine computation. There was von Neumann's "First Draft of a Report on the EDVAC," which contained the first detailed description of a universal stored-program computer. There was the three-part "Planning and Coding Problems for an Electronic Computing Instrument," written with Goldstine and Burks. Here was the notion of flow-charting, and machine-language programming. To foster the spread of knowledge, the authors intentionally refrained from copyrighting any of these papers, nor did they patent the machine itself. Von Neumann and his colleagues did the brainwork and experimental testing and then, in the traditional style of academic scientists, made their results freely available to any and all others.

Because the machine was new, and because Institute members didn't know how to exploit its capabilities, other Institute scientists didn't think much of the new electronic brain sitting in their own back yard. "There was never anything that we needed a lot of computing for," mathematician Deane Montgomery says. But above all there was this feeling that the genuine Platonic-heavenly scientist shouldn't be involved with *mechanisms*. "The snobs at our Institute," Freeman Dyson says, "could not tolerate having electrical engineers around them who sullied with their dirty hands the purity of our scholarly atmosphere."

There were people from the outside, of course, who were ready to pay for computer time, but Institute regulars regarded this as an absolute no-no. "We were not supposed to take outside contracts," Julian Bigelow says, "because they would be in some sense corrupting. And so we couldn't operate the machine. Finally it was taken over by Princeton University and operated by them for another three years."

In the late 1950s, after von Neumann died, faculty members and Institute trustees organized a committee to terminate the computer project. They held hearings on it, at the director's house, right in Oppenheimer's living room. "This was back when the Institute used to do things right," Harold Cherniss says. "Everything was informal."

The committee called people to testify, but it was all low-key, like the officers of a gentleman's club making a change in the bylaws. So Herman Goldstine came to Olden Manor and allowed as how the computer was no longer a research tool, that it was now ready for commercial development. Other people came in and told the same story, and at length the scholarly gentlemen decided to close down the whole project. "But we passed a more general motion," Harold Cherniss says. "It was a declaration to have no experimental science, no laboratories of any kind at the Institute." And so it has been ever since. The Platonic-heavenly fathers had triumphed. Or as Freeman Dyson puts it, "The snobs took revenge."

When von Neumann's unlamented computer was retired, in 1958, it went to the Smithsonian Institution, where it's now on public display. At the Institute for Advanced Study, by contrast, ECP room #1, where the computer was put together, is not treated as a historical site. No plaque or bust commemorates the birth of the stored program computer within. The room, at the end of a dark and lonely hallway, today houses the Institute's stationery supplies, and boxes of file folders, pads of paper, and inter-departmental mail envelopes reach almost to the ceiling. It might be thought poetically just that the room is also stacked high with that inescapable artifact of the computer revolution, data-processing paper, but the Institute's best monument to von Neumann is elsewhere, in the offices of John Milnor whose investigations of the Mandelbrot set would not be possible without the computer, and in the offices of Stephen Wolfram, whose computer simulations of cellular automata owe a lot to the mind and work of good-time Johnny.

At Los Alamos during the war years, it seemed that half of the world's top scientists could be found at the frequent dinner parties that took place at this secret city in the New Mexico mountains. At one of these affairs, when the conversation turned to the topic of extraterrestrials, to the possibility of intelligent life elsewhere in the universe, Enrico Fermi asked a famous question. If these extraterrestrials really existed, then he wanted to know just one thing: "Where are they?" The universe has existed for billions of years, he reasoned, long enough for many waves of extraterrestrial colonization to have reached planet Earth. The invaders should be here, all around us, perhaps even forcing us to do their bidding. But they're not. So, if they really and truly do exist, then . . . *Where are they*?

Dyed-in-the-wool believers in extraterrestrial intelligence have a perfectly good answer to this, of course. The ETs are home where they belong, just like we are. Frank Drake, for example, one of the founders of the SETI movement (the Search for Extraterrestrial Intelligence), says that aliens have decided that interstellar journeys are not worth the effort, and so they "are living comfortably and well in the environs of their own star."

Lately, though, the ET skeptics have rallied back with a new Fermi-type question: If there really are all these aliens spread throughout the universe, then: "*Where are their von Neumann machines*?" After all, a von Neumann machine is a self-reproducing universal constructor, a robot that makes copies of itself from whatever raw materials lie ready to hand—or to forceps, as the case may be. To reach other civilizations, all an intelligent species would have to do is to send out a small initial force of von Neumann machines, and these would sooner or later propagate, proliferate, and control the rest of space. "The key point," says mathematician Frank Tipler, "is that, once a von Neumann machine has been sent to another solar system, the entire resources of that solar system become available to the intelligent species which controls the von Neumann machine; all sorts of otherwise-too-expensive projects become possible." But we *don't* find extraterrestrial von Neumann machines invading downtown Dallas or Chicago, and so it's a good bet that there are no ETs out there after all.

It's a tribute to the care with which von Neumann described these "von Neumann machines," the self-reproducing robots that he conceived of, that today no mathematician or physicist doubts that they are theoretically possible. But why is this? Why are physicists and mathematicians, who are usually a conservative lot and not much given to daydreaming, why are they unanimously willing to entertain the proposition that brute machines, unfeeling creatures of meshing gears and cold steel, may somehow be able to engage in the process of self-reproduction?

The fact of the matter is that the idea of self-replicating machines is not really all that new to begin with. René Descartes, the seventeenth-century French mathematician and philosopher, maintained that animals are in effect no more than machines, and that people are just machines that have God-given souls in them. According to Descartes, there is nothing mystical or ineffable about humans or animals, at least so far as their *bodies* are concerned: bodies are simply physical systems which operate according to natural laws, just like everything else in the universe. Man's body, Descartes said, is "a machine which, having been made by the hands of God, is incomparably better arranged . . . than any of those which can be invented by man."

Descartes's view goes by a number of names: materialism, reductionism, mechanism, determinism. Underlying them all is the principle that everything in the universe—that is, every *physical* thing; imponderable entities like souls and spirits are another matter altogether—everything can be reduced to the operations of matter and motion. "All natural phenomena,"

Descartes said, "can be explained in this way; I therefore do not think any other principles in physics are either necessary or desirable."

Taken to its limits as a philosophy of science, the matter-and-motion view represents the extreme of optimism as regards man's ability to know nature. It's the view that once we understand the interplay of the atoms in their finest details, we'll know all there is to know, there will be nothing left over to elude the grasp of human reason, nothing mysterious lurking behind the phenomena, nothing hidden, nothing *left dangling*. There are no unseen spirits, occult life forces, or baffling animating vapors that can be comprehended only by intuition or a revelation from God on high. Mysticism dies a soggy, well-deserved death, and the universe is declared open for knowledge, utterly transparent to the human mind.

Call it immodest, arrogant, or what you will ("the epitome of hubris," perhaps?), such a viewpoint is basic to all of science. There's no way that a working scientist is going to spend a lifetime beating his brains out against the phenomena, trying to figure things out, if he's convinced deep down inside that nature, at bottom, is arbitrary and incomprehensible. "I should not want to be forced into abandoning strict causality," Einstein once said. "In that case, I would rather be a cobbler, or even an employee in a gaming-house, than a physicist."

Well then . . . if nature is open to our inspection, if there's nothing mysterious about the way things work, then why indeed shouldn't a machine be able to reproduce? Why shouldn't we be able to figure out how nature reproduces animals' bodies, and then go ahead and copy the process using man-made machines? Cells make cells, human bodies make human bodies, so why not . . . machines that make machines?

In June of 1948, while his electrical engineering staff was building a flesh-and-blood computer across the campus, John von Neumann gave a series of three lectures in Princeton on the subject of self-reproducing machines. (Here we have the basic ingredients of a mathematical-genius-run-amok scenario: With his subordinates putting together an electronic brain in the computer lab, the mad scientist himself, hair standing on end, lays plans for a race of self-replicating monsters that will take over the planet. It wasn't like this, of course . . . but on the other hand it was only a short while after von Neumann's lecture that another mathematician, Frank Tipler, had his visions of extraterrestrial von Neumann machines overrunning the galaxy.)

After the Princeton talks, von Neumann gave expanded versions of the lectures elsewhere. He wrote up some of his discoveries, but died before he put his full theory into final form. Later Arthur Burks, who had worked on the ENIAC and the I.A.S. computers, edited and completed Johnny's work on automata, and published the result as *Theory of Self-Reproducing Automata*. Almost certainly his most brilliant and original achievement in science, automata theory linked together von Neumann's work in logic, computers, and neurophysiology, and showed how the most basic property of life, reproduction, can be accomplished by simple mechanics.

The self-reproducing mechanisms that von Neumann invented are not creatures of the real world; they're abstractions: idealized, conceptual fictions that exist only in the imagination or on paper. Nevertheless, these abstractions contain the basic plan for machine self-reproduction.

"Now one has to be careful what one means by this," von Neumann said. "There is no question of producing matter out of nothing." Rather, he said, we have to think of machine replication along the lines of how animals, plants, and individual cells produce their offspring. They don't reproduce themselves *ex nihilo*, they utilize the raw materials in their environments, and the same must be true of machines. They'll have to have a ready supply of parts.

"Imagine," von Neumann said, "that there is a practically unlimited supply of these parts floating in a large container. One can then imagine an automaton functioning in the following manner. It is also floating around in this medium; its essential activity is to pick up parts and put them together, or, if aggregates of parts are found, to take them apart." This sea of machine parts is the mechanical equivalent of the earth's original primordial soup.

All earthly organisms have arisen through chains of evolutionary development that are quite random and accidental. There's nothing fated or necessary about the particular animals that exist here now: if the earth's initial conditions had been different, or if different mutations had occurred, then different species would exist from the ones that are here now. Von Neumann, by contrast, wanted to know what were the mechanisms that would have to be present for any evolution of any type to occur. He wanted to find the minimum necessary self-reproductive baggage, the Platonic archetype of genesis, as it were. And no miracles allowed: only matter in motion.

In his Institute lectures, von Neumann claimed that a self-replicating machine would have to have *at least* eight different kinds of parts, four for the brains, four for the brawn. The "brains" would be composed of organs that respond to different types of incoming stimuli. For example, if two stimuli occur together—for example if two necessary parts bump into the organism at the same time—the machine will have to know about this, and so it will have to have a sense organ that responds to and is aware of two or more simultaneous incoming messages. If the automaton—the robot—is being bombarded by all kinds of stimuli at the same time while it needs to be aware of only one of them, then it will have to have a faculty that can select out what it needs. In addition, it will need a sort of clock, an organ that can coordinate the actions of all the other parts.

As far as its body is concerned, a self-replicating machine would need an Archimedean point, something that will remain stationary with respect to something else. Call this a rigid member, or girder. Put one or more such together and you'll have the automaton's skeleton, its very bones. The skeleton could be internal, like man's, or external, like a lobster's, it makes no difference. The point is that it has to have some rigidity.

If the robot needs to put two things together that are floating around in the parts sea, it will have to have a fusing organ. By the same token, it will have to have something that will separate two or more things that are already connected together, so there will have to be a cutting organ as well. And then the robot will need something to make all these organs move, some "muscles."

There's no way to know, of course, what such a self-reproducing mechanism would look like, although it's easy enough to dream up any number of cute and cuddly little self-reproducers (see Figure 5).

As for the process of self-reproduction, suppose, von Neumann said, that floating around in the parts sea are two choice girders, and that the robot—also bobbing around in there—needs those girders in order to make a copy of itself. Suppose further that the girders bump into the robot's sensing organ, after which the robot connects them. It does this again and again, according to a plan, and soon there's a skeleton arising, where before there was only an assorted collection of parts. But where does the robot's "plan" come from?

Well, since the robot has sensing organs, it could learn the structure of some object—including itself—by touching it, and then recording the essentials of that structure in some kind of code. Later on, it could use that same code as a blueprint for making another version of the object in question. As for the code, von Neumann used a device of Alan Turing, who had discovered that any set of plans or instructions could be expressed in a binary notation, which is to say, a simple string of ones and zeroes. So von Neumann proposed that his automata use a binary code, and he went on to show how to make such a binary "tape" out of the rigid girders

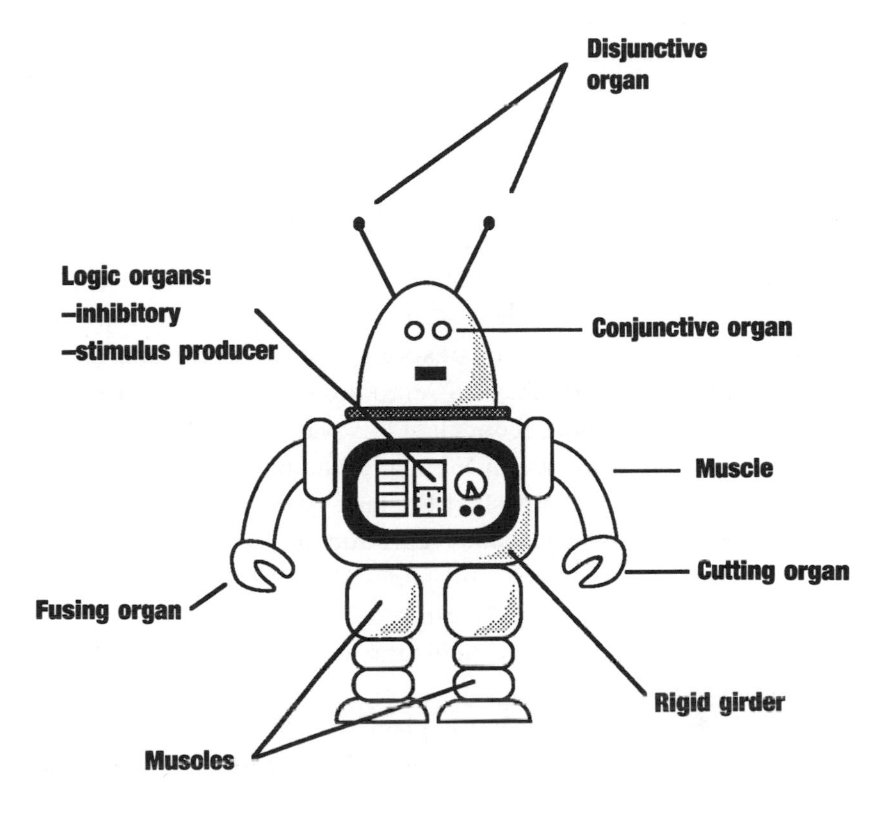

Figure 5 A kinematic automaton

floating around in the environment sea. The machine would take a bunch of girders and connect them up to form a sawtooth chain: /\/\/\/\/\. At each intersection it would then either insert a vertical girder, to stand for a 1, or leave the intersection blank, to stand for a 0. The code "word" 0 1 0 0 1 1, for example, would be represented by: /\/|\/\/\/|\/|\.

Once it had the structure of some object encoded in a "blueprint," it would be a simple matter for the automaton to duplicate the object that the code described. It would read the blueprint off the binary tape, select out the necessary elements from the surrounding parts sea, and then put them together according to the blueprint. The result would be a perfect copy of the object.

This would not be *self*-reproduction, of course, unless we imagine that the automaton had learned its own structure and encoded that into a blueprint, but there is no obstacle in principle to the robot's doing this.

Self-reproduction, then, can occur in the following way. First of all there are the givens: the automaton itself, the parts sea, and the blueprint. In addition, there's a mechanism that will make a copy of the blueprint, and, finally, there's a controller organ, which will direct all the operations and get them done in the proper sequence. Then the process begins. Following the blueprint, the robot takes what it needs from the parts sea—rigid beams, parts of muscles, random organs, and so on—joins some things together, and cuts other things apart. It arranges girders, organs, and whatnot, according to the blueprint, so that the structure of the object exactly matches its own structure. In the final step, the robot makes a copy of its own blueprint and attaches it to its offspring, leaving at the end an exact copy of the parent robot. Machine self-reproduction has occurred.

The strange thing about all of this was that, in figuring out how self-reproduction *had* to occur, von Neumann had hit upon the very way Mother Nature herself had accomplished the task. Von Neumann had worked out his abstract analysis of machine reproduction by December of 1949, four years prior to Francis Crick's and James Watson's explanation of the workings of the DNA molecule. It turns out that DNA molecules reproduce themselves precisely as von Neumann said any self-replicating machine would have to.

As Freeman Dyson has explained in his autobiography, *Disturbing the Universe*, "now every child learns in high school the biological identification of von Neumann's four components." The automaton itself, which does the work of replication, corresponds to the cell's ribosomes, the particles which translate the genetic information into molecules of protein. The copying mechanism, the part of the robot that makes a copy of its own blueprint, corresponds to RNA and DNA polymerase, the substances that combine nucleotides (rigid girders) into long chains of nucleic acids (the binary-tape blueprint). The controller, which directs the robot's operations, corresponds to the repressor and derepressor molecules that regulate gene development, causing different cells to develop in different ways. And finally there's the blueprint itself, containing in binary code the robot's own structure. This corresponds to the genetic materials themselves, DNA and RNA, which contain the genetic code.

"So far as we know," says Dyson, "the basic design of every microorganism larger than a virus is precisely as von Neumann said it should be."

Von Neumann even went on to explain how evolution can occur as machines reproduce. An increase in complexity can occur, he said, when an automaton's blueprint undergoes some kind of an alteration. Suppose, for example, that an automaton floating in the parts soup happens to bump into a rigid girder that's bobbing around next to it. If the girder hits the automaton just right, the collision may change a portion of its blueprint. Then, when it comes time to reproduce, the automaton will give birth, not to itself, but to a modified version of itself. A mutation will have occurred, and in this way fairly primitive automata—ones corresponding to the complexity of an amoeba, for example—can over the course of time give rise to other entitites that are comparatively sophisticated—like man. Artificial automata can evolve just as natural automata do, that is to say, just as animals have evolved. Complexity itself is the deciding factor. Below a certain minimum level of complexity, reproducing automata will degenerate into simpler mechanisms. But above that level, he says, "the phenomenon of synthesis, if properly arranged, can become explosive." A race of metal men could arise from a collection of nuts, bolts, and other parts bumping around in a primordial automata soup. John von Neumann, the Charles Darwin of the robots.

Von Neumann's analysis of "kinematic" or moving, three-dimensional automata, was not, by far, his last word on automata theory. In fact, it was only the beginning. Stanislaw Ulam, a colleague of von Neumann's at the Institute, and later on at Los Alamos, had once suggested that Johnny investigate an abstract, two-dimensional, checkerboard-like framework for automata. Ulam had used such a system of granular or "cellular" spaces to study the growth of crystals, and later von Neumann investigated whether an indefinitely large two-dimensional cellular space might not be enough of an environment for cellular automata self-reproduction. In the process of answering the question—in the affirmative—von Neumann created another whole branch of mathematics, cellular automata theory.

This was von Neumann's most abstract, high-up, Platonic-heavenly theory of them all. These logical entities, defined by mathematical functions alone, live, die, and reproduce in a vast, abstract, two-dimensional arrays of imaginary spaces. With the right functions programmed into them, the march of cellular automata across this abstract gridwork will approximate the evolution and development of the physical systems in nature.

As outlandish as the notion of self-reproducing two-dimensional cellular automata may have been, these entitites were quite acceptable to the other members of the Institute for Advanced Study. Cellular automata were, after all, only *abstractions*—Platonic archetypes—and these are what the Institute is all about. It turned out, however, that cellular automata do have a real-world significance. Just as von Neumann's three-dimensional robots elucidated the process of self-replication in living organisms, a later member of the Institute, Stephen Wolfram, would argue that von Neumann's automata have a wide and deep significance for the understanding of nature, that in fact they may be the same type of mathematical mechanisms that give rise to the complexity of the universe. The inner operations of cellular automata, he says, consititute a kind of "natural software." Whether or not he's right about this, what's absolutely certain is that Wolfram could never have advanced his argument without the aid of the instrument that John von Neumann had played a pivotal role in developing, the electronic digital computer.

Chapter 6

The Nim-Nim-Nim Man

The man who in a little more than two years will become the new director of the Institute for Advanced Study grabs onto a post, a riser supporting the rafters overhead, and steadies himself. "I must remain conscious," he thinks.

Tension is hardly the word for the mental state he's in. While he hangs onto the post and scans his flint-blue eyes over the tableau of technicians and instruments in front of him, the lanky, almost emaciated scientist scarcely breathes. The clock shows 5:29 in the morning, but it's as bright as day in here, inside the south 10,000-yard control bunker. It's because of the floodlights: they're shining onto an array of instrument panels crowded with knobs, switches, voltmeters, oscilloscopes, colored lights, relays, fuses, sequential timing mechanisms, firing circuits. All these are now bathed in a bright yellow glow so that movie cameras can record the movements of dials as they register the electronic messages of the apocalypse, the end of the world that is about to take place . . . in exactly 45 seconds.

The instruments are connected to wires that hang from the ceiling like the vines of a grape arbor. The wires all head off in the same direction, toward the shot tower at ground zero, some six miles in the distance. Atop the tower is the world's first atomic bomb, a malign-looking, six-foot ball of cables, implosion lenses, and detonators, known to all concerned as "the gadget." At this moment, it's the object of every man's attention: Sam Allison, as he does the countdown; Don Hornig, with his finger on the stop switch, ready to halt the proceedings in case some drunken rancher should happen to wander into the explosion area; George Kistiakowsky, explosives expert, who had bet the project director ten dollars against a month's salary that the gadget would work. Some twenty men here in the control room are waiting for the final moments of eternity to tick away, for the coming of an event that will border on the supernatural.

Sam Allison's voice, counting off the last, long seconds, booms out across Trinity Site: "Ten . . . nine . . . eight . . ." It echoes over to the north 10,000-yard observation shelter where Princeton physicist Robert Wilson worries that, because of the winds, the mushroom cloud is going to waft directly overhead. The count reverberates across Base Camp, where I.I. Rabi and Enrico Fermi are stretched out on the desert sand, facing away from the blast site. On Compania Hill, twenty miles to the northwest, Richard Feynman, Hans Bethe, and Edward Teller listen by radio: ". . . seven . . . six . . . five . . . four . . ."

In the main control bunker, as he marks off the final seconds of a dying age— ". . . three . . . two . . . one . . ."—Sam Allison realizes that the blast may send an electrical impulse up into the microphone he's holding, electrocuting him. He drops the mike like a hot potato and screams at the top of his lungs, *"Zero!"*

There's a lull in which heartbeats, the world, time itself, seems to stop . . . And then the instrument panel is flooded with a ghastly pulsating radiance, as if a lightning bolt had struck just outside the doorway . . . and in that blinding instant, as the thunder of doom shudders through the air and the ground and a molten fireball rises up off the desert floor, some lines of poetry, from the *Bhagavad Gita*, flash through the head of J. Robert Oppenheimer, the man hanging onto the post . . .

I am become Death,
The destroyer of worlds.

Indeed, at this moment J. Robert Oppenheimer is the unlikeliest man in the world to be the next director of the Institute for Advanced Study.

His role as "father of the atom bomb" was only one of many that Oppenheimer played during his lifetime, but it's the role he played the best. It was Oppenheimer, after all, who managed to get the world's top scientists to come to an unknown mesa in the New Mexico desert, without telling them exactly where they were going, nor for how long, nor precisely what they would be doing when they got there. But he made the prospect sound irresistible. "The projected laboratory as described by Oppenheimer sounded romantic," said Robert Wilson, much later. "And it was romantic. Everything to do with it was to be clothed in deepest secrecy. We were all to join the army and then disappear to a mountain-top laboratory in New Mexico."

"Even the location was vague," physicist John Manley remembers. "I had learned from Oppie, who knew the country, that the site was in the 'Hamos' Mountains. But I could not find any such range on any available map. With no exposure to Spanish, how could I know he said 'Jemez'!"

Oppie drew scientists away from their classrooms, from their cyclotrons, from their laboratories. They came from Princeton, Harvard, M.I.T., the University of Chicago. When these easterners, who were used to staring at the black streets and grey office buildings of New York and Boston, got off the train at Lamy, New Mexico, a desert whistle-stop bursting at the seams with some fifty people, they couldn't believe their eyes. *There was nothing out there.* There were some low hills in the distance, a clump of trees here and there, and there was this little Mexican-style adobe station, but that was it. You looked down the tracks one way and then you looked down them the other way, and you could see for fifty, sixty, seventy-five miles . . . but *there was just nothing there.*

A G.I. car or truck would show up to meet you, though, and after a while you were being driven through some of the most spectacular, exciting scenery you had ever seen . . . past weird rock formations and the sweet-smelling sage . . . through flat desert that stretched for fifty or a hundred miles in any direction. Way off on the horizon, shimmering through a tremulous purple haze was this range of snow-capped mountains that seemed to just hang there, motionless. Above it all there was this incredible big throbbing blue sky that seemed to arch and vault to infinity. It was mind-boggling, exhilarating . . . it just about knocked your eyes out.

Soon you were passing some low red cliffs, and then some higher ones, and then the car was actually heading up into the mountains, up to the secret mountaintop lab, climbing up a dirt road that switched back and forth, with straight-down frightening drop-offs on either side. "The air was clear and mild," I.I. Rabi remembers, "the Sangre de Christo Mountains were distinct and sharp, the mesa on the other side—lovely! And the ride up on the old road, somewhat hair raising but very interesting, the old bridge, and then, of course, the Indians; we certainly seemed to enter a new world, a mystic world."

Up on the mountaintop J. Robert Oppenheimer reigned as omniscient scientist, omnipotent administrator, and head mystic. To him the canyons and piñons of Los Alamos were a second home. His two loves were physics and New Mexico, and he used to spend his summers at a ranch there which he named Perro Caliente—"hot dog"—which is what he said the first time he laid eyes on the place. There Oppie played another one of his great roles—that of New Mexico cowboy. "He eventually explored a large part of those mountains," says his friend Francis Fergusson, "and he probably knew more about them than almost anybody else. He would just get on his horse and put a chocolate bar in his pocket, and he'd be gone for a day or two."

"I'd never been on a horse in my life," says physicist Robert Serber. "They gave us maps and they set us off on these three-day trips over the mountain passes at 12,500 feet. You went out on the absolute minimum of equipment: a bottle of whiskey and graham crackers—and oats for the horse."

Later, at Los Alamos, Oppie ruled over a hand-picked, elite group of scientists—"the biggest collection of eggheads ever," General Groves called them—they were the chosen few whose mission it was to save Western civilization from the metaphysical madmen Hitler, Mussolini, and Hirohito.

It was 1943. Only a few years earlier, in 1938, Otto Hahn and Fritz Strassman had discovered that the uranium atom could be split. The next year, Frédéric Joliot-Curie published a paper in *Nature,* showing that 3 or 4 neutrons were emitted in every fission: a chain reaction—an *explosive* chain reaction—was possible. As incredible as it seemed, a lump of matter no bigger than a grapefruit could, under the right conditions, be turned into molten kinetic energy. Mother Nature was turning out to have some pretty strange propensities: one of her true sons, one of the naturally occurring elements—number 92 on the periodic table—could be transformed into . . . a bomb.

At least that was the theory. Uranium had been discovered by Martin Klaproth way back in 1789, and for the last 150 years not one person had ever seen any of it just . . . explode. But Oppenheimer and his cadre of physicists, they were going to take this bare theoretical possibility and turn it into a reality, into a device. Before they were finished, they were going to take a lump of that pure silvery matter and make it . . . blow up on demand.

The conversion of matter into energy that took place on a desert morning in 1945 took science past a point of no return. No longer a disembodied, ephemeral probing after nature's secrets, science became a bloodied instrument of war and death, and Oppenheimer, perhaps more than any other single individual, was responsible for the change. Yet two years after the explosion at Trinity Site, the father of the atom bomb would ascend to the throne of the One True Platonic Heaven, to become director of the Institute for Advanced Study for the next nineteen years.

As the world outside went off to war, so did a few of the members of the Institute for Advanced Study, much to the dismay of Abraham Flexner. Although he was no longer the director, Flexner was still an Institute trustee, and one thing he was firm about was that his faculty members ought to remain on the job. After all, when they had signed on, they pledged themselves to give their full time and attention to the Institute, and Flexner saw no reason why a mere war should change matters. Nonetheless, almost everyone at the Institute, including the current director, Frank Aydelotte, did war work of some type, and a few of them actually went to Europe, as did mathematician James Alexander, who journeyed to England. He worked at Bomber Command Headquarters, trying to improve the accuracy of bombing strikes.

More usually, however, a faculty member remained at the Institute and did what he could from there. Professor Erwin Panofsky, the art historian, prepared maps and lists of information about cultural monuments in Germany, this too for the use of American bombers. Many of the scientists, including Marston Morse, Oswald Veblen, and John von Neumann did consulting work for the Army, and shuttled back and forth between the Institute and the Aberdeen Proving Grounds, in Maryland, and, in von Neumann's case, also to Los Alamos. The only faculty member who didn't get much involved in this was Albert Einstein.

Not that he was averse to doing so. It was Einstein, of course, who had, together with and at the urging of Leo Szilard, written the famous 1939 letter to Roosevelt advising the President that "extremely powerful bombs of a new type" could be made out of fissionable materials such as uranium, and asking that he "speed up the experimental work" which was then being conducted. But Einstein was quite willing to do more in order to help build the bomb. In December of 1941, Vannevar Bush, director of the Office of Scientific Research and Development, asked for Einstein's help with certain problems pertaining to the gaseous diffusion method of separating U-235 from other uranium isotopes. Einstein was only too happy to comply, and he offered Bush his advice in a handwritten manuscript, adding that he would be glad to do more, but that he needed further information.

Bush, though, wouldn't give Einstein more information, for the German physicist was regarded as a security risk. "I wish very much that I could place the whole thing before him and take him fully into confidence," Bush told Aydelotte, "but this is utterly impossible in view of the attitude of people here in Washington who have studied into his whole history."

Einstein therefore spent the war years in Fuld Hall, working, according to an Institute publication, on "possible means for the construction of a unified theory of relativity."

By the time the war ended, the Institute had a total of ninety-two members and faculty. Four of the original faculty members, including Einstein, had formally "retired," but, as the Institute was careful to point out in its annual *Bulletin*, "In every case this retirement is merely technical. All four of the members of the faculty are continuing their scholarly work and their activities are duly recorded in the chapters devoted to the appropriate School." Their retirement was so "technical," in fact, that Einstein and Veblen—neither of whom were at all interested in retiring—were kept on at full salary and continued to attend faculty meetings just as though they were regular members. The major significant change in their status consisted in their now being listed in the Institute *Bulletin* as "Professors Emeritus."

The Institute was now on its own private campus, and faculty members had settled into their brand new offices in Fuld Hall. A radio was installed in the common room, where it was used for Saturday night dances. During the week, faculty wives served teas. Outside Fuld Hall, in the front meadow, a bowling green was put in. Evidently, Institute members were not much interested in bowling, for the lane was soon gone. The members spent more of their time in the Institute's Ford car, which made daily trips back and forth between Fuld Hall and Fine Hall on the Princeton University campus.

Behind the bucolic exterior, the usual infighting and backstabbing continued on, most of it focused, also as usual, upon the director. In 1945, Frank Aydelotte would turn sixty-five, which according to bylaws, meant that he would retire from the directorship. But people had mixed feelings about this. On the one hand, it was fresh in most people's minds how Abraham Flexner had become a dictator in his old age, and no one wanted to go through anything like that again, and so from this angle Aydelotte's retirement looked good. On the other hand, Aydelotte was a gentle, peaceloving Quaker who mostly let people do what they wanted, and from this angle any change in the status quo looked bad.

Aydelotte was no great scholar, but faculty members didn't hold this against him in the least. Einstein even used to joke that a good director should be "a little stupid," so that he wouldn't spend all his time hatching forward-looking plans, schemes, and new institutional policy directives. If there was anything that members didn't want to hear about, it was "new directions" or anything of the sort. As Oswald Veblen put it in a letter to the trustees, "There is a strong feeling among the faculty that it would be a mistake at the present time to bring to the Institute a new Director who might come with a preconceived policy." The feeling was that Aydelotte should stay on for a while, despite the bylaws, and despite Aydelotte's own desire to relinquish the directorship.

This led to Faculty Mutiny Number Two.

In the fall of 1944 the Institute faculty called a special meeting and passed a resolution that said, among other things, that "the present Director knows how to work with scholars"—as opposed, of course, to Abraham Flexner, who didn't—"and as a result there exists a spirit of harmony and effective cooperation in the Institute which has been reflected in the substantial achievements in the last five years." (This was not merely braggadocio. Notwithstanding the interruptions caused by the war, the publications of members during 1943/44 exceeded in volume those for any previous year.) The faculty therefore wished that the present director would remain, but if that were impossible then they had another suggestion. They advanced the truly shocking notion that perhaps the faculty could manage quite well on its own, that the Institute could function without a director. (Later on, Oppenheimer suggested—in all seriousness—that the Institute could probably function *without a faculty*. Perhaps the truest Platonic Heaven of all would have *no people*, just a collection of wraiths.)

This heretical idea had been fostered, although quite unwittingly, by Aydelotte himself, who had given the faculty more of a say in running the Institute than Flexner ever had. It was symbolic of this that, when he wrote up the text that was to be published in the annual Institute *Bulletin*, Aydelotte spelled the word "faculty" with a capital "F," whereas Flexner had always used the lowercase letter. Also, in December of 1945, Aydelotte had accepted President Truman's invitation to be a member of the Joint Anglo-American Commission on Palestine. Aydelotte was away from the Institute for five months as a consequence, but the place somehow kept together in his absence, governing itself by means of a standing committee of professors who made all the necessary decisions. That this worked out as well as it did gave the faculty confidence that such an arrangement might well be made permanent.

Veblen, who often seemed to be running the Institute behind the scenes, like a ghost, now suggested that the Institute have a "Rector rather than a Director," and that this new position, which would have far less power than the directorship used to, would rotate from person to person every two years.

Nothing ever came of this. After the usual bickering in committee meetings, faculty meetings, and so forth, it turned out that Aydelotte would indeed stay on until a successor could be found. Most important of all, the Institute would continue to have a director and not, as Veblen had proposed, a "rector."

The trustees formed a special committee to recommend a successor to Aydelotte, and the faculty did likewise. The faculty committee consisted of three professors: James Alexander, of the School of Mathematics, Edward Mead Earle, of the School of Economics and Politics, and Erwin Panofsky, of the School of Humanistic Studies. Early in 1946, the faculty committee sent to the rest of the faculty a list of seven candidates, arranged in the following order:

Dr. J. Robert Oppenheimer, physicist, University of California,
Dr. Detlev Bronk, physiologist and physicist, of Philadelphia,
Dr. Harlow Shapley, astronomer, director of the Harvard Observatory,
Mr. (formerly Major General) Frederick Osborn,
Professor Edward S. Mason, economist, Harvard University,
T.C. Blegen, historian, and dean of the Graduate Schools, University of Minnesota, and,
Professor E. Harris Harbison, historian, Princeton University.

Three weeks later, the faculty committee added two more names to the list, making for a total of nine nominees for the directorship:

Dr. Henry E. Sigerist, historian of medicine, Johns Hopkins, and
Mr. (formerly Rear Admiral) Lewis L. Strauss, trustee, Institute for Advanced Study.

Shortly afterward the whole faculty met for a business luncheon to discuss the candidates. They reduced the list to five names, which they then listed alphabetically: Blegen, Bronk, Mason, Oppenheimer, and Strauss. The trustee committee, for its part, added the name of Linus Pauling to the faculty's new, pared-down list, but did not object to any of the candidates. For some unknown reason, the name of Lewis Strauss was left on the list, this in spite of the fact that the man had virtually no academic credentials at all.

Strauss—who pronounced his name "Straws," the way they did back in old Virginia, where he came from—was a self-made man in every sense of the word. He'd started off as a wholesale shoe salesman in the deep south, where he used to drag two trunks full of shoe samples into the coal mining towns of West Virginia and North Carolina. He never went to college, but he did give himself an education, of sorts. On his days off he used to read law, Latin, and the sciences, but he mostly learned at the school of hard knocks, in the business world.

Strauss went to Wall Street where he made a small fortune as a partner of the international banking firm of Kuhn Loeb & Company. He went into the Navy in 1941 and later became Rear Admiral. The man had a definite knack for making the best of whatever circumstance he was in, and had a near-miraculous instinct for managing money. It was this last attribute, indeed, which had caused the Institute for Advanced Study to elect him a trustee in 1945, and which now led Oswald Veblen to propose him for director in what one professor described as "a long and facetious speech."

Equally surprising to some, however, was the name of J. Robert Oppenheimer. The notion of "the father of the atom bomb" as the director of the One True Platonic Heaven was utterly incongruous, and Professor Benjamin Meritt, the Greek epigrapher, wasn't speaking merely for himself when at a faculty meeting he stated his hope that no one "too intimately associated with the atomic bomb would be appointed." But beyond the bomb issue was the quite separate fact that Oppenheimer had already been considered for an appointment as a professor and had been passed over in favor of someone else. That a man who was not good enough to be a faculty member could now be considered seriously for the directorship was another incongruity. Nevertheless, there it was.

Oppenheimer was a scientist and maker of The Bomb, but he was also a poet and a writer of short stories. He was writing poems by the age of ten or twelve, and later, at Harvard, he had one published in the avant-garde literary magazine *Hound and Horn*. He studied philosophy, literature, languages. He knew eight languages and used to read Plato's dialogues in the original Greek, and the *Bhagavad Gita*, the ancient Hindu epic, in Sanskrit. Once when a couple of his friends—Fritz Houtermans and

George Uhlenbeck—were reading Dante in Italian, Oppie felt left out and so he spent a month or so teaching himself the language, after which he joined in with their readings. In the Netherlands he was asked to teach a physics seminar, and did this happily enough, giving the talk in Dutch. "I don't think it was very good Dutch," he admitted.

Oppenheimer had gotten interested in science when at an early age—about five or seven or so—his grandfather gave him a collection of minerals. "From then on I became, in a completely childish way, an ardent mineral collector and I had, by the time I was through, quite a fine collection." Later he developed an interest in chemistry, and by the time he got to Harvard he was able to go through their four-year science program in three years, graduating with a degree in chemistry *summa cum laude* in 1925.

At the time, the best physics wasn't being done in America, but in Europe, so Oppie left for England, where he hoped to study with Ernest Rutherford, at Cambridge. But "Rutherford wouldn't have me . . . My credentials were peculiar and not impressive, certainly not impressive to a man with Rutherford's common sense," and so Oppie worked instead with J.J. Thomson. Soon enough, though, Oppie decided that he wasn't meant to be an experimentalist. Much of the lab work seemed pointless: "I think my intention was to get the art of making thin films, perhaps to postpone the more difficult question of what I would do with them when I got them," he said of one lab session. "I did make some beryllium films . . . but I will not try to tell you the miseries of evaporating beryllium onto collodion and then getting rid of the collodion and so on."

Oppenheimer turned into a theoretical physicist at Göttingen, where he studied with Max Born. He and Born collaborated on a paper in quantum theory, and later, when Oppie received his Ph.D. degree, Born told him: "It's all right for you to leave, but I cannot. You have left me too much homework."

Oppie then took himself to wherever the important work in physics was being done, and so he went to the University of Leiden to study with Paul Ehrenfest, to the University of Utrecht to work with Hendrik Kramers, and to the Federal Institute of Technology in Zurich (Einstein's alma mater) to work with Wolfgang Pauli. By the time he returned to the States in 1929, Oppenheimer had already published sixteen scientific papers, six of them in German, all of them on the hottest subject in theoretical physics, quantum theory. He decided to spread his new knowledge, establishing himself in California, which was then quite in the hinterlands. "I thought I'd like to go to Berkeley because it was a desert," he said. "There was no theoretical physics and I thought it would be nice to start something. I also thought it would be dangerous because I'd be too far out of touch so I kept the connection with Caltech."

So Oppie took on joint appointments, spending one semester at the University of California at Berkeley, the other at the California Institute of Technology, in Pasadena. Some of his students liked him so much that they followed him from place to place. "We thought nothing of giving up our houses or apartments in Berkeley," Robert Serber remembers, "confident that we could find a garden cottage in Pasadena for twenty-five dollars a month."

It was in California that Oppenheimer began to cast his spell over a whole generation of physicists. He was not a particularly good classroom lecturer, at least in the beginning. A chain-smoking stick figure, Oppie would pace back and forth across the room, flailing his spindly arms around, sometimes speaking in tones that were so low that they verged on the inaudible. He'd hesitate and grope for words, often making a succession of halting "nim-nim-nim" sounds (sometimes it was "hunh, hunh") to mark time between sentences and paragraphs: ". . . and that's the function of the Dirac constant here . . . . nim-nim-nim. . . But the picture changes when you consider . . . hunh, hunh . . ." And so on.

Soon his students were copying Oppie's style, right down to the chain smoking, the fingernail biting, the gesturing, the blue shirts, even the nims. Wolfgang Pauli took to calling him "the nim-nim-nim man," and when a European physicist once stated in Pauli's presence that there was "no real physics" being done in the United States, Pauli replied, "Oh? You mean you haven't heard of Oppenheimer and his nim-nim-nim boys?"

Oppenheimer and his nim-nim-nim boys worked in particle physics for about ten years, but suddenly, in the fall of 1938, they were doing astrophysics, and thinking about the death throes of stars. Oppie wrote three papers in quick succession, each of them with a different graduate student, and each of them projecting a more violent end for the more massive stars. The last of the three, written in 1939, contains Oppenheimer's prediction that some stars will collapse down to black holes.

The idea of a black hole, a celestial body whose gravity is so strong that even light can't escape from it, was not original with Oppenheimer. It went all the way back to the eighteenth century, to Pierre Simon de Laplace, the French astronomer and mathematician who, in his *Exposition du Système du Monde* broadcast the message that the universe is nothing but a big, all-encompassing mechanism, a giant clockworks in the sky. The world, and everything in it, he said, is so rigidly controlled and predictable that, if it were possible to know, in minutest detail, its state at any one moment, then you could deduce its state at any other moment of the past or future.

Gravity affected all material bodies, and since light, as both Newton and Laplace believed, was composed of tiny corpuscles of matter, it was also subject to gravitational pull. If a star was big enough, Laplace reasoned, then its emitted starlight would be held back by its gravity, perhaps even to the extent that the star would be invisible to us on earth: "A luminous star of the same density as the earth, and whose diameter was two hundred and fifty times greater than that of the sun, would not, because of its attraction, allow any of its rays to arrive at us; it is therefore possible that the largest luminous bodies of the universe may, through this cause, be invisible."

In the twentieth century, Laplace's speculations were confirmed by the German astronomer Karl Schwarzschild, who applied Einstein's general relativity equations to a hypothetical case in which all of a star's mass is concentrated at a single dimensionless point. Schwarzschild found that, if the mass at that point were sufficiently condensed, then the space around it would close in on itself so that light would be unable to escape. The range at which this closing-off effect will occur will vary with the mass involved. If earth's mass were shrunk to such a point, then this distance—now known as the Schwarzschild radius—would be roughly one centimeter, but larger masses would make their effects present across larger volumes of space.

Schwarzschild showed that something like Laplace's invisible bodies could exist within the framework of general relativity, but the problem was that Schwarzschild's dimensionless points were mathematical abstractions, not objects in the real world. It was Oppenheimer who applied general relativity to real-world phenomena and showed that the black hole was in fact a genuine physical possibility. He began by analyzing the collapse of stars.

A star dies when it runs out of its hydrogen fuel, and as it cools down, the process of stellar collapse begins. What a given star will end up as depends upon its initial mass. If its mass is less than 1.4 that of the sun, then it will contract to become a white dwarf, a star made up of superdense matter, about 1 percent the size of the sun. Bigger stars undergo more spectacular, violent contractions, ending up as neutron stars, exploding into supernovas, and losing in a few moments an amount of heat equivalent to the total energy it had radiated over its previous billions of years of life. What's left behind is a tiny cinder of matter, a neutron core. Composed of tightly packed neutrons, a cubic inch of the stuff would weigh 10 billion tons.

For a while astronomers had believed that a collapsing star would stabilize at the neutron-core stage and not contract any further, but in his 1938 paper "On Massive Neutron Cores," written together with his Berkeley grad student, George M. Volkoff, Oppenheimer calculated that, if the neutron core were big enough (if it had a mass greater than .7 that of the sun), then once the collapse began, it would just keep on going. (Oppie's first paper, written with Robert Serber, discussed the relationship between a star's mass and the possibility of its collapsing to form a neutron core.)

Six months later, Oppie was writing a new paper "On Continued Gravitational Attraction," this time with Hartland Snyder. The two now calculated that the star's nucleus would continue to shrink, its gravity growing ever stronger and progressively cutting off the emission of light until, at the bitter end, the star would blink out like a candle. As the authors expressed it, the star "tends to close itself off from any communication with a distant observer; only its gravitational field persists." Their calculations provided the best evidence then available that black holes are real.

Oppenheimer had been considered for a professorship in the winter of 1945, about five months before the first A-bomb was exploded at Trinity Site. The faculty members compared Oppie against Wolfgang Pauli, whom they were also thinking of hiring. At the time, Pauli was already in residence at the Institute, although as a temporary member. Albert Einstein and Herman Weyl were asked to write a memorandum in which they evaluated the two men. Although their Institute colleagues scarcely needed to hear it, Einstein and Weyl began by expounding upon the general superiority of pure theory over crass experimentation.

"The School of Mathematics," they wrote, "is of the unanimous opinion that theoretical physics not only should continue to form a part of its scientific activities, but should even be reinforced. The entire history of physics since Galileo bears witness to the importance of the function of the theoretical physicist, from whom the basic theoretical ideas originate. *A priori* construction in physics is as essential as empirical facts. Of course the theorist must have contact with the discoveries and findings of experimental physics, but it is enough that laboratories exist in the civilization in which he lives; it is by no means necessary that he be associated with a laboratory at the place where he works."

That out of the way, the authors went on to state their clear preference for Wolfgang Pauli. The scientific accomplishments of the latter, they said, were more important than those of Oppenheimer. "Certainly Oppenheimer has made no contributions to physics of such fundamental nature as Pauli's exclusion principle and analysis of electronic spin. . . . Pauli's command of the mathematical apparatus is, and probably always will be, far the greater."

Einstein and Weyl did praise Oppenheimer for his "excellent administrative work" during the war, and for his having "founded the largest school of theoretical physics in this country." They noted that his interests were broad, that he habitually surrounds himself with a brilliant social circle, and his students are very enthusiastic about him. "It may be," they cautioned, "that he is somewhat too dominant, and [that] his students tend to be smaller editions of Oppenheimer."

In the spring of 1945, therefore, the Institute's School of Mathematics formally offered a permanent appointment to Wolfgang Pauli. Pauli, however, was undecided about accepting, and in the end he stayed on at the Institute only until 1946, whereupon he returned to Zurich. Before he left, the School suddenly moved to recommend that Oppenheimer be offered a professorship of theoretical physics, and directed Einstein and von Neumann—neither of whom really cared for the man—to prepare his resumé for consideration by the faculty and trustees.

Oppie, for his part, was never in any hurry to come to the Institute. He had visited the place back in 1935, and was not impressed. "Princeton is a madhouse: its solipsistic luminaries shining in separate & helpless desolation," he wrote to his brother Frank at the time. The Institute's Herman Weyl had offered him a position on that occasion, but Robert had turned it down. "I could be of absolutely no use at such a place," he told his brother, "but it took a lot of conversation & arm waving to get Weyl to take a *no*."

Now it was the Institute's turn to drag its heels. A year after Einstein and Weyl were supposed to have prepared his vita, the School of Mathematics had still not made any formal offer to Oppenheimer. They decided instead to make an overture to Julian Schwinger, who had worked under Oppenheimer at Berkeley. Schwinger, however, wasn't interested in coming to the Institute.

The School then tried to make an unprecedented deal with Richard Feynman. "Somehow they knew my feelings about the Institute," Feynman said later, "how it's too theoretical; how there's not enough *real* activity and challenge. So they write, 'We appreciate that you have a considerable interest in experiments and in teaching, so we have made special arrangements to create a special type of professorship, if you wish: half professor at Princeton University, and half at the Institute.'"

But to Richard Feynman, this was incredible. It was an honor, of course, but it was utterly too much to live up to, "an absurdity," he thought. "I laughed at it while I was shaving, thinking about it." So, like Oppenheimer and Pauli and Schwinger before him, Richard Feynman passed up a position at the Institute for Advanced Study.

Robert Oppenheimer, meanwhile, had resigned as director of the Los Alamos laboratory, and was wondering what to do next. The natural thing seemed to go back to teaching, and Oppie received no small number of offers from the top universities and research institutes, including Caltech and Berkeley again, but also Columbia, and Harvard. In the end, Oppie returned to his old posts in California. But things were not the same. "The charm went out of teaching after the great change in the war," Oppie remembered much later. "I did teach at Caltech and Berkeley and for one thing I was always called away and distracted because I was thinking about other things, but I actually don't think I ever taught well after the war."

Which is not, of course, any surprise. Mere teaching would have to be quite anticlimactic, quite a letdown, compared to being the director of the Los Alamos Laboratory. At Los Alamos you were going head-to-head with the greatest names in physics every day of the week, and you had this incredible *power* in your hands. When you saw that energy—which was your very own creation—when you saw it shine like the sun and send a million tons of rock and rubble lofting up into the night sky . . . well, that was a lot more exciting than teaching.

But if there was no going back to Los Alamos, the next best thing was maybe to be in charge of another major research institution, another place whose sole purpose and entire justification was to gather together the most brilliant minds of the age—the prima donnas—and let them discover the secrets of the universe. While the Institute for Advanced Study might not be too attractive as a place to be a professor, matters stood differently when it came to being *director*. Directing the Institute might be at least a pale reflection of how it had been back at Los Alamos. How fortunate, then, that the Institute happened to be looking for a new director, and how convenient that Robert Oppenheimer's name should happen to be at the top of the list.

In a way, Oppie made perfect sense from the Institute's point of view. The place was top-heavy with mathematicians and it needed a few good physicists to balance things out and revitalize everyone, and who better to do this than the great miracle-worker from Los Alamos? He seemed to be on a personal, first-name basis with every last one of the world's top physicists. He was a brilliant intellect, a legendary administrator. He got people to do things that were beyond them. Besides, Oppenheimer was not *just* a physicist, he was a master of language, he knew poetry, literature, philosophy, all those things that the Institute stood for and held sacred. He was as close to a universal genius as one could hope for in modern times. As one of the Institute regulars explained it much later, "Hell, this is a mecca for intellectuals and we're reading in the *New York Times* every day that Oppenheimer was the greatest intellectual in the world. Of course we wanted him—then."

So they offered him the job. In the fall of 1946, trustee Lewis Strauss flew out to California to visit the Radiation Laboratory at Berkeley. He was met at the airport by Oppenheimer and Ernest O. Lawrence, who was head of the Rad Lab. Strauss took Oppie over to the concrete ramp in front of the hangars, and offered him the directorship of the Institute for Advanced Study. The salary would be $20,000, plus retirement at age 65 with a pension of $12,000. Additionally, the director would live in Olden Manor, the eighteen-room mansion on the Institute grounds, rent-free, with servants and all the rest of it. Oppenheimer said he'd like time to consider the offer, but at least he was interested.

Oppie took his time, all right, for by the following spring he still hadn't made up his mind. He loved California, particularly Berkeley, and wasn't convinced that the Paradise for Scholars in Princeton was all that it was cracked up to be. "I regard it as a very open question whether the Institute is an important place, and whether my coming will be of benefit," he said shortly afterward.

But one April night as he and his wife Kitty were driving across the bridge from San Francisco to Oakland, they heard on the car radio that J. Robert Oppenheimer had accepted the directorship of the prestigious

Institute for Advanced Study in Princeton. "Well, let that decide it, then," he said.

When he arrived at the Institute in the fall of 1946, it was clear from the start that Oppenheimer was going to be far different from his predecessors. In the director's large, first-floor offices at one end of Fuld Hall, Frank Aydelotte had hung up on the walls a collection of prints showing scenes of life at Oxford. Aydelotte had been an English professor, and had gotten his literature degree from Oxford, in 1908. But these wistful academic prints were not to Oppie's liking, and so he took them down and put up a blackboard that ran the length of the wall. Most of the time it was covered with equations.

And then there was the matter of the office safe. Formerly it had been used to store legal papers and confidential documents, but when Oppie came to the Institute he brought with him a package of top-secret papers that dated back to the Los Alamos days. For these, a mere safe wasn't enough, and so out in the hallway, day and night, around the clock, an armed guard stood watch over the atomic secrets. People feared that Oppie would bring laboratories to the Institute. He didn't; he brought guns.

In fact, he brought worse than guns. In June of 1951, a committee of bomb experts assembled at the Institute to discuss plans for the "Super," the thermonuclear, or hydrogen bomb. Edward Teller was here, as were Hans Bethe, Norris Bradbury, who was the new head of the Los Alamos Lab, Enrico Fermi, John Wheeler, and of course the Institute's own resident bomb expert John von Neumann ("one of the best weapons men in the world," in the estimation of Gordon Dean, chairman of the Atomic Energy Commission, who was also in attendance). Oppenheimer had earlier opposed the H-bomb's development, but now, as he listened to the new ideas of Teller and Stan Ulam, Oppenheimer changed his mind. "When you see something that is technically sweet you go ahead and do it," he explained, "and you argue about what to do about it only after you have had your technical success."

In the beginning, one of the things that Oppie found attractive about the Institute was that it was only a few hours from Washington, where by now he had acquired an aura of wisdom and authority that was quite metaphysical in its proportions. Heads of government wanted to know what to do about The Bomb, how to use this new instrument which science had given them, and here was Oppenheimer who . . . well, he *was* The Bomb. Who would know better than he how to use it? He became an administrator and consultant. He was elected chairman of the GAC—the General Advisory Committee—of the Atomic Energy Commission; he was a member of the Department of Defense's Research and Development Board; he worked for the Air Force's Science Advisory Committee of the Office of Defense Mobilization. Oppie traveled across the United States and abroad, to give lectures, to attend meetings, to chair panel discussions. He became the great Pooh-Bah of atomic energy, and he loved it.

Some Institute members worried about Oppie's being an "absentee director," but he wasn't really neglecting the place. It was just that the directorship didn't require all of his time, far from it. When he left Los Alamos there were six thousand people doing his bidding, and now he was lord and master over a mere hundred or so, all of whom had their own private projects. They didn't need "directing," didn't need leadership the way the atomic scientists did. From the start Oppie had insisted that he would be an administrator only half of the time; the other time he'd spend doing physics. It was written down in his letter of appointment. "By vote of the Trustees," the letter said, "you were also appointed Professor of Physics, to serve concurrently as such with your service as Director, this appointment having been made in accordance with your wish to continue in your chosen field."

This was a first for the Institute, but it worked out well enough. Oppenheimer put his stamp on the Institute in a way that no other director has, before or since. He emphasized physics, of course, but he also emphasized youth, much to the delight of some of the old-timers who had fears about the place becoming an old-age home. Einstein, for one, worried that the hiring of older men "might make of the Institute an institution." So Oppie brought in a bunch of young physicists: Abraham Pais, Freeman Dyson, T.D. Lee, and C.N. Yang, among others. Some of these appointments were almost magical. A few years after Yang and Lee arrived, the two of them won the Nobel prize for physics. They had overturned one of nature's "laws," the law of conservation of parity.

Chen Ning Yang and Tsung-Dao Lee were in their mid-twenties when they first came to the Institute for Advanced Study. They were born in China and had attended the same college in Kunming, the Southwest Associated University, which was then a collection of tin-roofed shacks housing classrooms and separate thatched-roofed huts housing dormitories. The classroom windows were frequently shattered by air raids, for this was wartime, and the fact that anyone got educated there—much less went on to win the Nobel prize—is a testimony to the brute power of mind over matter.

Yang and Lee came to the United States on fellowships to attend the University of Chicago, where Lee would write a doctoral dissertation on white dwarf stars under Enrico Fermi, while Yang would do work in nuclear physics under the guidance of Edward Teller. It was at Chicago that Yang and Lee started to collaborate, producing a series of thirty-two joint papers in physics. The two were together again at the Institute from 1951 to 1953, then Lee took a position at Columbia University in New York, while Yang remained in Princeton. But they were only fifty miles away from each other, and they made it a point to meet once a week to talk about problems in the physics business.

At this stage in the development of elementary-particle physics, there were some twenty known particles, including electrons, protons, neutrons, neutrinos, positrons, mysterious "V-particles" (later known as "strange particles"), and an entire family of entities known as mesons, whose members seemed to proliferate like rabbits. Things got so bad that one physics textbook included a chapter entitled "Particles We Might Do Without." As for understanding the whys and wherefores behind these fast-multiplying subatomic entities, physicists were at a loss. In a 1953 article in *Scientific American*, Freeman Dyson painted a bleak picture indeed: "Nobody has had any success in classifying the known particles, or in predicting the properties of the unknown ones. Nobody understands why such and such particles exist, why they have the particular masses that are observed or why some of them strongly interact and some do not."

But as they found new particles, physicists rushed to make sure that they obeyed all the "old" laws—like conservation of energy—and also the exotic "new" laws that they were then coming up with—laws such as conservation of charge conjugation and baryon number. These conservation principles have to do with what happens in particle interactions, as for example when one given entity disintegrates into others. A few particles, such as protons, electrons, and neutrinos, are thought not to decay at all; that is to say, they are "stable." Most others do decay, or are unstable, some having lifetimes that are exceedingly short, often only billionths of a second long. In these interactions, certain characteristics of the system change, while others remain the same. The ones that remain the same are said to be "conserved," and one of the most fundamental of these is electric charge. In every particle decay, the total net charge of the system involved is the same both before and after the interaction. A neutron, for example, which is electrically neutral, decays into a proton and an electron, whose opposite charges cancel each other out, leaving at the end of the reaction the same electrical charge that there had been back at the beginning.

In 1956, Yang and Lee were investigating the case of the K-meson (or kaon). Mesons had been discovered back in 1938 by Seth Neddermeyer and Carl Anderson, two of Oppenheimer's colleagues at Caltech. The K-meson is unstable, and breaks down into another type of meson, called a pion (pi-meson), in about a microsecond. But some K-mesons were observed to decay into two pions, others into three. This in itself was not too troubling, because it was known on the basis of both theory and experiment that one and the same particle could have two different decay modes. What was troubling about the kaon decay was that the parity of the two sets of decay products was different. That shouldn't happen. Parity should be conserved just like electric charge or anything else.

Parity is the general property of mirror-symmetry, the fact that an object and its mirror-image obey the same laws and are functionally equivalent in every way. For example, if you stand in front of a mirror and throw a ball from hand to hand, parity means that the ball in the mirror responds just as it does in the real world: it will describe a parabolic arc in both cases. As Frank Yang put it in his Nobel acceptance speech (he had renamed himself Frank in honor of Benjamin Franklin, whose great fan he was), "the laws of physics have always shown complete symmetry between left and right."

Parity was thought to exist on the elementary particle level as well as in macroscopic nature, although the way it manifests itself on the micro level is a bit different. In the case of particles, parity is designated by plus and minus signs, corresponding to what happens when spatial variables pertaining to particle wave functions are reversed. If the wave function is not changed, the particle is said to have *even parity* (designated by a plus sign); if the wave function changes when its orientation in space is reversed, then it is said to have *odd parity* (designated by a minus).

What happened with the K-mesons is that sometimes they decayed into particles of even parity, at other times into odd-parity particles. This was puzzling because there was no other difference between the original mesons: they had almost the same mass, the same electrical charge, the same everything. To the physicists of the period, this was unintelligible: the law of parity conservation demanded that identical particles have identical decays.

To account for this strange behavior, physicists had just a few available alternatives. One was to say that they were in fact observing the behavior of two different K-mesons, which they proceeded to name the tau (τ) and the theta (θ), respectively. The trouble with this suggestion, however, is that no one could find any intrinsic difference between the two particles. It was a distinction without a difference.

The other alternative was to say that parity conservation, at least in this case, simply wasn't a fact of nature after all. The trouble with this suggestion was that to deny parity conservation in any one case meant that the "law" of conservation of parity really wasn't a law. Plainly, denying one of the hitherto accepted laws of nature is not a matter that any physicist takes lightly. Nevertheless, this is exactly what Yang and Lee decided to do.

They published a paper called "Question of Parity Conservation in Weak Interactions" in the *Physical Review*. (Yang and Lee had originally entitled the piece, "Is Parity Conserved in Weak Interactions?" but the editor of the *Review*, Samuel Goudsmit, would tolerate no question marks in titles.) In their paper, Yang and Lee summarized the reasoning which led them to question the whole idea of parity conservation: "Recent experimental data," they wrote, "indicate closely identical masses and lifetimes of the theta and tau mesons. On the other hand, analysis of the decay products of tau strongly suggest on the grounds of angular momentum and parity conservation that the theta and tau are not the same particle. This poses a rather puzzling situation. . . . One way out of the difficulty is to assume that parity is not strictly conserved, so that theta and tau are two different decay modes of the same particle. . . ." To find out whether parity is in fact conserved, Yang and Lee suggested that an experiment be done "to determine whether weak interactions differentiate the right from the left." At Columbia University, Madame Chien-Shiung Wu read the paper of Yang and Lee and decided to do the experiment.

The idea behind it was to see whether electron radiation will occur symmetrically in all directions. If parity held for weak interactions—the interactions involved in particle decays and radiations—then electrons ought to be emitted from a radioactive substance equally in all directions. If parity was not conserved, more electrons would be emitted in one direction than in another. It would be as if nature, for some reason, had some kind of directional preference. This didn't seem likely.

The basic problem was how to line up atomic nuclei so that they'd all be pointing in the same direction to begin with. Wu did this by taking a piece of material containing cobalt—specifically, the isotope ^{60}CO—and cooling it to within a few tenths of a degree of absolute zero (−273° C). When the substance was at this low temperature, she then applied a strong magnetic field to the sample, and this had the effect of aligning all the cobalt nuclei in the same direction. With all of them headed the same way, the nuclei should emit electrons equally in any direction as long as parity was conserved. If parity was not conserved by the underlying interaction, then more electrons would be sent out one way than the other.

When the moment of truth was at hand, it turned out that the cobalt emitted radiation preferentially. Nature, for reasons still unknown, was like an invisible hand at work directing electrons where she wanted them to go. The result was clear: Yang and Lee were right. Parity had broken down. For their disproof of parity as a law of nature, Yang and Lee were jointly awarded the Nobel prize for physics in 1957.

At the time they won the prize, Yang had been a professor at the Institute for two years, whereas Lee was a visiting member. Later, Lee joined him there as a professor, and for a while the twin Nobelists were a happy sight for everyone. Oppenheimer used to comment that it made him proud just to watch Yang and Lee walk the Institute grounds. Here at last was a collaboration that seemed to work.

But as a place where people can actually work together the Institute has always had a rather bad reputation. "There's not a terribly well-developed tradition of cooperation at the Institute for Advanced Study," says Murray Gell-Mann. "The original idea was that it was a place for thinkers, individual great thinkers, to go and think. This is a relatively sterile mode for modern times. I mean it might be all right for Einstein, who was out of tune with the physics of his time anyway. He didn't need anybody to talk with because nobody was interested in his stuff and he wasn't interested in what anybody else was doing. But for other people, cooperation seems to be what produces results."

The Institute's lack of cooperative spirit is traceable to the prima donna factor, to the me-me-me mental attitude that sometimes affects its great minds, often without their even being aware of it. Toward his sixtieth birthday, some of Frank Yang's friends offered to put together a festschrift in his honor, a collection of original essays written in tribute to his work in physics. So they put the idea to him. "I thought about it," Yang says, "and concluded that a collection of some of my papers with Commentaries might be more interesting."

So in 1983 publisher W.H. Freeman brought out Yang's *Selected Papers 1945-1980 with Commentary*. At Columbia University, T.D. Lee read

Yang's version of their famous alliance and blanched. It wasn't as he remembered it at all. Shortly thereafter, Lee wrote up his own version of events and circulated it around privately to his friends, then published it in volume three of *T.D. Lee: Selected Papers,* which was brought out by Birkhäuser in 1986. If their experience is at all characteristic of what happens when two bounteous egos try to pool their talents, it may help to explain why there are not more collaborations at the Institute.

Here is Yang on his relationship with Lee at the University of Chicago: "Being older and several years ahead of him in my graduate studies, I tried to help him in every way. He later became Fermi's thesis student, but he turned to me for guidance and advice, and I was in effect his teacher in physics for those Chicago years. . . . I was like an older brother to him. . . . I bent over backward to attempt to help him in his career. . . ." And so forth.

In 1952, the two physicists, both of them now temporary members at the Institute, wrote a long, two-part paper on gas-liquid phase transitions. Egos were apparently starting to blossom out, for it suddenly became a matter of vital concern which of their names should appear first on the title page. As Lee tells it, Yang wanted it to be "C.N. Yang and T.D. Lee." Since Lee was four years younger than Yang, he at first acceded to this. Later, he checked some other jointly written physics papers and discovered that age did not necessarily dictate the order in which the names were listed. So Lee wanted it to be "T.D. Lee and C.N. Yang."

Fortunately, they had written the paper in two parts, and so it turned out that each could have it his own way: it was "C.N. Yang and T.D. Lee" in part one, "T.D. Lee and C.N. Yang" in part two. Both parts were published in the same issue of *Physical Review,* back to back.

At the Institute, Albert Einstein read the paper and decided he wanted to talk to the authors. According to Yang, Einstein sent for *him*: "Einstein sent Bruria Kaufman, who was then his assistant, to ask me to see him. I went with her to his office, and he expressed great interest in the paper. . . . Unfortunately I did not get very much out of that conversation, the most extensive one I had with Einstein, since I had difficulty understanding him. He spoke very softly, and I found it difficult to concentrate on his words, being quite overwhelmed by the nearness of a great physicist whom I had admired for so long."

According to Lee, Einstein sent for *them*: "Einstein asked his assistant, Bruria Kaufman, to see if he could talk with Yang and me. . . . We went to Einstein's office. . . . Our answers pleased him. . . . The whole conversation was quite extensive and lasted a long time."

Yang and Lee did not collaborate again for a while, until 1956, when they wrote up their findings on the nonconservation of parity. Yang's version is that he wrote the paper himself: "I showed the manuscript to Lee, who made a few minor changes. I then signed our names in alphabetical order. Briefly, I considered putting my name first, but decided against it, both because I disliked name ordering and because I wanted to help Lee along in his career." Lee's version: "I remember that the writing was a joint effort; indeed, as usual in our collaborations, we debated almost continuously the wording and nuances of our presentation. There were several versions of the paper, revised over a period of time."

Ego was a problem again in Sweden. "I was in Princeton when the news of the 1957 Nobel physics prize came," Lee says. "During the month of November my wife, Jeanette, and I prepared for the trip to Sweden. Yang and I also had to write our lectures and speeches. In the course of discussing what to do, Yang asked if our actual accepting of the prize during the ceremony could be done in age order. I was surprised, but acquiesced."

Five years later, in 1962, a profile of Yang and Lee, both of whom were now full professors at the Institute, appeared in *New Yorker* magazine. The two physicists were sent galley proofs of the article in advance of publication, and each made corrections. But Lee says that Yang now wanted his own name to come first in three places in the article: in the title itself, at the point in the text where the Nobel prize is announced, and in the description of the award ceremony. Yang also wanted his wife's name to precede the name of Lee's wife in the text, on account of the fact that Yang's wife was a year older.

"I told him he was being silly," Lee says.

On April 18th, 1962, according to Lee's recollection, Yang came into Lee's office at the Institute and asked for even more name juggling in the *New Yorker* piece. Appalled, Lee wondered aloud if their collaboration could continue. Yang "then became very emotional and started to cry," Lee says, "saying he wanted very much to work with me. I felt embarrassed and helpless, and talked to him gently for a long time. In the end we agreed we would stop collaborating at least for a limited time, and then decide." Yang's version: "It was an emotion-draining experience, with cathartic senses of liberation."

When the *New Yorker* article finally came out a month later, Lee's name came first in the title—"A Question of Parity: T.D. Lee and C.N.

Yang"—whereas Yang's name came first at the initial mention of their winning the Nobel prize. Otherwise their names appeared in a more or less random ordering.

That spring, Lee resigned from the Institute to take up a position at Columbia University in New York. There is a rumor that Oppenheimer went up to Columbia to see if he could persuade Lee to return, but failed. In 1966, Yang also left the Institute to go to the State University of New York at Stony Brook.

Twenty years later, in November of 1986, T.D. Lee celebrated his sixtieth birthday with a big party at Columbia University. Many important physicists showed up for the occasion, including I.I. Rabi, Sidney Drell, Bruno Zumino, and Freeman Dyson. For better or for worse, Frank Yang was not there.

Although he worked at physics on a half-time basis, Robert Oppenheimer never had much of a scientific output while he was at the Institute. People have given a host of explanations for this. Oppie was burnt out; he was past the age when physicists can be expected to do any good work; he was too involved in Washington politics; all of the above. Everyone seems to be agreed, likewise, that Oppie never lived up to his full potential as a scientist, however much he may have exceeded himself as Los Alamos administrator. For this too there is no lack of explanations.

"It seems to me," says I.I. Rabi, "that in some respects Oppenheimer was overeducated in those fields which lie outside the scientific tradition, such as his interest in religion, in the Hindu religion in particular, which resulted in a feeling for the mystery of the universe that surrounded him almost like a fog. . . . Some may call it a lack of faith, but in my opinion it was more a turning away from the hard, crude methods of theoretical physics into a mystical realm of broad intuition."

Others point to Oppie's impatience with details—"His physics was good, but his arithmetic awful," says Robert Serber—and to his lack of persistence. "He didn't have *Sitzfleisch*," says Murray Gell-Mann. "Perseverance, the Germans call it *Sitzfleisch*, 'sitting flesh,' when you sit on a chair. As far as I know he never wrote a long paper or did a long calculation, anything of that kind. He didn't have patience for that; his own work consisted of little *aperçu*, but quite brilliant ones. But he inspired other people to do things, and his influence was fantastic. He taught one or two generations of American theorists about modern physics, quantum mechanics and field theory."

People have complained that Oppie was fast, but not original, that he would grasp your thoughts and finish them for you, but that they were still *your* thoughts, and not his own. Still others have pointed to his occasional duplicity, his capacity for speaking out of both sides of his mouth. One of his most famous utterances had to do with The Bomb: "In some sort of crude sense which no vulgarity, no humor, no overstatement can quite extinguish, the physicists have known sin; and this is a knowledge which they cannot lose." (This public hand-wringing incensed many of his colleagues, including Oppie's old teacher at Harvard, Percy Bridgman. "If anybody should feel guilty," said Bridgman, "it's God. He put the facts there.") And at a meeting with President Truman after the war Oppie confessed, "Mr. President, I feel I have blood on my hands." (Truman, hearing this, is supposed to have taken a handkerchief out of his pocket, saying, "Would you like to wipe them?") But toward the end of his life, Oppenheimer was painting quite a different picture of events. "What I have never done," he said, "is to express regret for doing what I did and could at Los Alamos. In fact, on varied and recurrent occasions, I have reaffirmed my sense that, with all the black and white, that was something I did not regret."

It was Oppenheimer's mental quickness, combined with a selective liberty with the truth, that led to his well-publicized downfall. Two weeks before Christmas, on December 14, 1953, Oppenheimer talked on the telephone with Lewis Strauss, the man who had, six years earlier, offered Oppenheimer the directorship of the Institute. Strauss was now chairman of the Atomic Energy Commission. Oppenheimer himself was a consultant to the AEC, and Strauss wanted to talk to Oppie about his security clearance. So Oppenheimer went down to Washington to see Strauss. What Strauss had to tell him made his head spin.

The fact was that the president himself, Dwight D. Eisenhower, had placed a "blank wall" between Oppenheimer and classified atomic energy matters. All remaining classified documents would be yanked from Oppenheimer's safe at the Institute, and Oppie's security clearance was suspended forthwith. The reason was that, in the words of William L. Borden, who had been an official of the Joint Congressional Committee on Atomic Energy, "more probably than not J. Robert Oppenheimer is an agent of the Soviet Union."

The crux of the matter was of course Oppie's communist connections: his brother Frank was a communist, and so was Frank's wife. For that matter, so was Oppie's *own* wife—or at least she had been, back in the early days. Oppie had even employed communists, lots of left-leaning physicists, at Los Alamos. And there was the matter of some untruths he had told to the Los Alamos security officers a long time ago. But the strangest part of it was that the government had known all this back *before* Los Alamos and yet . . . *they had cleared him*. They had let him build The Bomb. So why were they rehashing all of this now, ten years after the fact?

The conventional explanation is that Oppie had publicly embarrassed his one-time supporter, Lewis Strauss, and that Strauss, to get even, would see Oppie damned to hell.

On June 13, 1949, Oppenheimer testified before the Joint Congressional Committee on Atomic Energy, which was considering the Norwegian government's request to be shipped a millicurie of iron 59. They wanted to use the isotope for the purpose of monitoring the behavior of molten steel in the smelting process. Strauss, a fire-eating anticommunist, was leery of exporting radioactive materials of whatever kind, but when he found out that one of the men on the Norwegian research team had some communist leanings, well . . . that was it. No isotopes for the Norwegians as far as Strauss was concerned.

With Strauss present in the committee room, Oppenheimer took the witness stand to testify in favor of sending the isotopes. He was asked if they could be used for military purposes. "No one can force me to say you cannot use these isotopes for atomic energy," Oppie said. "You can use a shovel for atomic energy. In fact you do."

There were a few Haw Haws in the room, and Strauss's teeth began to grind. "You can use a bottle of beer for atomic energy. In fact you do." Heh, heh, heh. "But to get some perspective," Oppie continued, "the fact is that during the war and after the war these materials have played no significant part and in my knowledge no part at all." Meanwhile, Strauss was getting red in the face and was staring daggers at the lordly atomic scientist up there making fun of him.

Senator Knowland asked Oppie, "Is it not true, Doctor, that the over-all national defense of a country rests on more than secret military development alone?"

"Of course it does," Oppie said. "My own rating of isotopes in this broad sense is that they are far less important than electronic devices, but far more important than, let us say, vitamins."

That was it for Strauss. A few years later, after William Borden had branded Oppie "an agent of the Soviet Union," Strauss saw this as his golden opportunity to get back at Oppenheimer. Strauss arranged for the FBI to shadow Oppie wherever he went, to intercept and open his mail, and, according to a recently declassified FBI report, to plant a bug in Olden Manor, Oppenheimer's home at the Institute for Advanced Study. "At the specific request of Admiral Lewis L. Strauss, Chairman, Atomic Energy Commission," says the FBI file, "a technical surveillance was installed at the residence of Dr. J. Robert Oppenheimer at Princeton, New Jersey on 1-1-54."

Oppenheimer had brought not merely guns and H-bomb conferences to the Institute, but wiretaps, FBI agents, the works.

In the end, Strauss won. Oppenheimer was branded a security risk and officially stripped of his clearance. The big security safe at the Institute, and its 24-hour-a-day guards, were removed. Strauss, though, wasn't yet finished with J. Robert Oppenheimer. He also wanted Oppie fired from the Institute. This the trustees and faculty adamantly refused to do. In June of 1954, the faculty got together and drafted a statement supporting Oppenheimer. It read, in part, "For seven years now [Dr. Oppenheimer] has with inspired devotion directed the work at the Institute for Advanced Study, for which he has proved himself singularly well suited by the unique combination of his personality, his broad scientific interests and his acute scholarship. We are proud to give public expression at this time to our loyal appreciation of the many benefits that we all derive from our association with him in this capacity." The statement was signed by all twenty-six permanent members and emeritus professors, including Julian Bigelow, Freeman Dyson, Albert Einstein, Kurt Gödel, Herman Goldstine, Abraham Pais, Atle Selberg, Oswald Veblen, John von Neumann, Herman Weyl, and C.N. Yang.

Oppenheimer, though, was by now a beaten man. He lost his self-confidence and arrogance, which may have finished him off as a scientist. And he lost his consulting posts in Washington, which means that he spent more time at the Institute. Some faculty members saw this as a good thing. "So far as I am concerned," Freeman Dyson has said, "he was a better director after his public humiliation than he had been before. He spent less time in Washington and more time at the Institute. . . . He became more relaxed and more attentive to our day-to-day problems. He was able to get back to doing what he liked best—reading, thinking, and talking about physics."

But Oppie's humiliation took its toll. Both Oppie and his wife Kitty were big drinkers, and after his defeat, things seemed to get worse. "I can remember a typical evening at that house," one visitor recalls. "You would sit in the kitchen, just gossiping and drinking with not a thing to eat. Then about ten o'clock Kitty would throw some eggs and chili into a pan and, with all that drink, that's all you had." It got to the point where mathematician Deane Montgomery went around calling Oppie's place "Bourbon Manor." Oppie, for his part, called Montgomery "the most arrogant, bullheaded son-of-a-bitch I ever met."

Evaluations of Oppenheimer's nineteen-year tenure as director are more or less divided along party lines, with the physicists remembering him fondly for having placed them at the center of attention, and the humanists admiring him for his vast erudition outside the sciences. On one occasion Oppenheimer was asked for some advice by Lansing V. Hammond, who was working for the Commonwealth Fund, a program set up by the British government to send bright and promising young British scholars to study in the United States. Hammond was awash in applications and called on Oppie for help with his physics applicants.

Oppenheimer handled the physics applicants pretty quickly, but then asked to see the others, even those in the humanities.

"Umm . . . indigenous American music," Oppie said, looking at an application form. "Roy Harris is just the person for him; he'll take an interest in his program . . . Roy was at Stanford last year, but he's just moved to Peabody Teacher's College in Nashville . . . Social psychology; he gives Michigan as his first choice . . . Ummm . . . he wants a general, overall experience; at Michigan he's likely to be put on a team and would learn a great deal about one aspect. I'd suggest looking into Vanderbilt; smaller numbers; he'd have a better opportunity of getting what he wants."

"Symbolic logic," Oppie said, turning to another application. "That's Harvard, Princeton, Chicago, or Berkeley; let's see where he wants to put the emphasis. . . . Ha! Your field; eighteenth-century English lit. Yale is an obvious choice; but don't rule out Bate at Harvard; he's a youngster but a person to be reckoned with."

And so it went for the next hour or so. Of some sixty applications, Oppie was unable to help with only two or three. For the others . . . well, he seemed to be just about omniscient. No wonder the humanists liked him.

Oppie's main enemies at the Institute were the mathematicians, who saw him packing the place with physicists, and paying scant attention to their own subject. "He was out to humiliate mathematicians," says André Weil. "Oppenheimer was a wholly frustrated personality, and his amusement was to make people quarrel with each other. I've seen him do it. He loved to have people at the Institute quarrel with each other. He was frustrated essentially because he wanted to be Niels Bohr or Albert Einstein, and he knew he wasn't."

Oppenheimer did, however, have quite a high opinion of himself. In 1963, after about fifteen years as director, Oppenheimer talked to an interviewer about the strengths and weaknesses of the Institute for Advanced Study. "As a place for permanent members we are not good for more than a very few people," Oppie said. "In the world of theoretical physics the number of people who are better off here for life than elsewhere is not very large, and they do typically spend six months of the year here and six months of the year elsewhere. Dyson taught all last year, Yang is teaching all this year. Dyson will be at La Jolla next year, Yang is at Brookhaven much of the time even when he's here; and in these ways they fulfill those needs without which they couldn't survive."

The strength of the Institute, he said, is as a refuge, as an intellectual hotel. "For a postdoctoral fellow or for someone who is a practicing physicist that has lacked contacts or has become a little too enmeshed in teaching and responsibility at home, a year or, more typically, six months of the year with nothing to do but physics and plenty of physicists to talk to is wonderful. That's what this place is about."

But then he went on to muse that perhaps the Institute doesn't need a faculty after all, this in view of the fact that Oppie could *almost* perform their main function himself, which is to guide and advise the temporary members: "If I were multi-valent enough, knowledgeable enough, and vigorous enough," Oppie said, "we wouldn't need a faculty, if I could talk to all members about their problems in all fields."

If anyone could have brought this off, it would have been J. Robert Oppenheimer.

Extremes of Vision

Chapter 7

Hubble, Bubble Toil and Trouble

Franz Moehn is without a doubt one of the most important and revered figures at the Institute for Advanced Study. He's been here only since 1979, a comparatively short time (people stay there sometimes for decades). And, even though at an Institute where members tend to separate themselves off from each other quite rigidly according to disciplinary lines, everyone seems to know him, everyone has been affected by his work. Franz Moehn is, of course, the chef.

Moehn was born in Wittlich, Germany, in the Moselle Valley region. His father ran a hotel there, called the Zum Rebstock, which is where Moehn learned how to cook. Later, he came to the United States to serve an apprenticeship with the Sheraton Hotels, after which he came to the Institute, where now he's got his own private, cookbook-filled office down in the dining hall. Thanks mainly to his fine cooking, the dining hall has turned out to be the *de facto* social center and focal point of the Institute, the place where all its members gather together every working day. They don't break disciplinary ranks, of course, but at least they're in the same room.

The kitchen, off limits to members, is a cook's paradise of white-tiled walls and stainless steel appliances. There's a meat locker, in which hang freshly killed pheasants—their feathers still on; there's a large, walk-in freezer, a separate vegetable cold room, a baking area, plus a group of ranges, convection steamers, ovens, and smokers. Moehn, who's got some bluefish in there smoking, is at the moment putting a caramel glaze on some fifty small white ramekins prior to filling them with flan. These are for the paella dinner tomorrow night. The dining hall serves coffee and rolls every morning, lunch every weekday, and dinner every Wednesday and Friday. These dinners are *events*, a typical menu being: appetizer (perhaps steamed shrimp), salmis of guinea hen, wild rice, gratin of salsify, salad, dessert.

Or: appetizer (perhaps melon with prosciutto), scampi on spinach risotto, frisee with celeriac, zabaglione. Or: appetizer, poached skate in noisette butter, green beans forestière, salad, dessert. Or: appetizer, roast pheasant, spinach with shitake mushrooms, salad, dessert.

Meals such as these wouldn't be complete without wine, and on each dinner night the menu lists a small selection of suggested wines from the Institute's cellars. When Moehn first came to the Institute he found that their wine collection left something to be desired, and so he went off on a buying spree, with the full support of director Harry Woolf, who likes the finer things in life. (Oppenheimer used to walk the couple of blocks between Olden Manor and his office in Fuld Hall; Harry Woolf drives his Audi.) Moehn now buys for a decade in advance, and the Bordeaux he's purchased have yet to be placed on the wine lists. As a result of his efforts the Institute currently boasts a collection of some 5,000 bottles.

The lunches here are like the dinners, although on a reduced scale. Entrées include breast of chicken teriyaki, the ever-popular veal fricassee over puff pastry, Flemish carbonades, moussaka, Swedish pork roast, and of course fresh fish according to market every Friday. For lighter appetites there's a huge selection of cold dishes and salads, plus sandwiches ranging from tongue on rye to chicken salad on pita bread. Wines are available at lunch too, together with foreign and domestic beers, and there's always a whole raft of desserts: rhubarb pie, raspberry torte, fresh custard in ramekins, whipped cream over everything. Espresso coffee is there to top things off, ninety cents a cup. If you're an Institute member you pay for it all with a flick of your Institute charge card. People who come here oftentimes find themselves putting on weight; they owe it all to Franz Moehn.

The dining room itself is a large rectangle skylit on three sides, clerestory fashion, with the fourth wall completely glassed in so that members look out upon a garden. There are flowers out there according to season—daffodils right now, for it's the first day of April—a stand of tall birch trees, and a waterfall which flows out into a long blue pool. You can eat in the garden when the weather's good, but if you stay inside you'll still find fresh flowers in small vases on the tables. On the walls hang mosaics, from the ancient city of Seleucia, in Mesopotamia. Right next to one of them is a modern abstract painting.

What with all these attractions it's no wonder that the Institute regulars start heading off for the lunchroom as soon as the clock creeps past 12. Inside, of course, they congregate according to discipline, the astrophysicists usually bunching together at one of the long tables at the north end, the particle physicists at a couple of tables farther south. The mathematicians seem to break up further according to age group, the younger members here, the older faculty—Selberg, Weil, Montgomery—going off separately over there. Freeman Dyson usually eats with the astros—the "little gang of astrophysicists," as he calls them—even though his office is in with the particle physicists. Dyson, though, is a case apart, and virtually uncategorizable. His work ranges from molecular biology to theoretical mathematics to particle physics to building nuclear reactors and spaceships. He likes the astrophysicists, though, for their sense of humor.

Every Tuesday, Dyson's gang of astrophysicists take themselves off to a separate area called the Board Room, where they have their weekly working luncheon. So attractive is this event to astrophysicists in the area that it brings them in not only from the Institute, but also from the university and from nearby Bell Laboratories.

It's 12:30 now and the Board Room is filling up. The center of the place is taken up by a number of wooden dining tables arranged in the shape of a U, around which, in Breuer-style cane-backed chairs, sit the Princeton-area astrophysicists. There are about fifty of them altogether, digging into their main courses now, waiting for John Bahcall to give the word. Bahcall, at the head of the U along with today's speakers, is currently executive officer of the Institute's School of Natural Sciences, a position that rotates from year to year. He's also the chief astrophysicist-in-residence, specializing in the solar neutrino problem, about which he's writing a book. Gossip has it that he's in line to be the Institute's next director ("I'm not on the list of candidates, nor would I allow myself to be considered," he says), but right now he's the genial host for the Tuesday working luncheon of the Princeton-area astrophysicists. Bahcall, in a blue-and-red-striped sweater, runs things low-key, which is very much his style.

He clinks a teaspoon against his water glass and turns to Don Schneider. "Well, Don, why don't you tell us about the exciting new results that we've been discussing at the Institute all morning."

Schneider, one of the younger members, holds a five-year Institute appointment—which means that he's at the upper limits of post-doctoral brilliance. He had already broken the news during morning coffee, back in building E, but now it's time to share it with the wider group of scientists who have come here to commune with others of their species.

Don clears his throat. He's a bit nervous, but who can blame him? After all he's going to announce that just this morning, in the wee hours, he's discovered the most distant object in the known universe.

"I have some data here," Don says, holding a sheaf of papers, "that pertain to a new quasar. It has a redshift of four point one. . . ."

Four point one! Everyone in the audience knows what that means. In astrophysical lingo distances of the very farthest objects—quasars, remote galaxies—are expressed not in light years or megaparsecs but in units of redshift. For a while, back in the 1970s, about 650 quasars—quasi-stellar objects—had been identified, having redshifts up to about 3.5. That meant they were billions of light years away from our galaxy. Then some others were found even farther away, with redshifts of 3.6 or 3.7, that's right at the edge of creation. But now, *4.1!* This is too sudden, almost too much to believe. A few people in the audience even stop chewing for a moment, a sign of really big news. But Don has even more to tell them.

"The object appears as part of a gravitational lens," Schneider says. "We've measured the absorption lines and there doesn't seem to be much doubt about it. We were just very lucky."

A gravitational lens! Incredible! Gravitational lenses are galaxies that split the light rays coming from objects much farther off, objects like quasars. Some of the quasar's light goes around one side of the galaxy, some of it careens around to the other side so that you get this double image—one on either side of the galaxy—of one and the same object. Einstein had predicted gravitational lenses a long time ago, and then, very recently, a few of them were actually found in the heavens. But they are exceedingly rare phenomena, and with good reason. The quasar, galaxy, and observer—that is, we on earth—all have to be lined up just right, or the lensing effect won't occur. But here's Don Schneider announcing not only that he's discovered the furthest object in the universe but that it's in the middle of a gravitational lens to boot. This is a bit much!

But Schneider hands out his charts to the audience, and there's no disbelieving the data. Wavelength plotted against energy flux, the graph looks like a distorted view of lower Manhattan, with sharp peaks and valleys, and one very sharp spike, looking like the World Trade Center. That's the quasar, with its record-breaking redshift.

The whole room is abuzz. People are talking to each other a mile a minute, and John Bahcall has the devil of a time moderating the question period. They want to know everything: Where's the object located? What's its coordinates? What's the exact time the observations were taken? But Don answers them all, every last one . . . until it's clear that the thing has gone far enough, and he brings it all to a close. There's another speaker to be heard from, poor fellow. Schneider is going to be one tough act to follow.

Indeed. Some people are even now getting the drift, an inkling of what's actually been going on here. A redshift of *four point one*, and today is *April first*. Can this be? . . . Oh, Jesus! It *must* be. And in fact, yes, it is! It's all . . . an *April Fool's joke!* Don Schneider has just pulled off the coup of the decade, getting the combined astrophysical brains of Princeton University, Bell Labs, and the Institute for Advanced Study to believe that in the space of a few hours in the morning, at an Institute with absolutely no observing facilities whatsoever, not even so much as a pair of binoculars, he's discovered the world's farthest object smack in the middle of a gravitational lens.

Schneider can't believe it. He thought the whole thing would be a flop, that the august astronomers of Princeton would see through his tricked up data and fake charts in a minute. He's been having a tough time trying to keep a straight face—and so has John Bahcall, who's been in on the gag from the start. But a few people still don't get it. At the end of the lunch they come up to congratulate Schneider, to tell him what a great job he's done, really good work. They want to shake his hand. They're quite serious. The discovery is much too good to be true, but they believe it all anyway.

But Don lets them know the awful truth. He's been talking nonsense all this time.

If they knew Don Schneider better, they might have been prepared. After all, he's the one with a picture on his office door of Ricardo Montalban as Khan in *Star Trek: the Wrath of Khan*, and underneath the picture the caption says "Don Schneider." And then there's this poster, just the type you might expect to find on the door of any concerned, conscientious, politically aware academic. It's very convincing, quite authentic-looking, quite like the real thing. But when you actually get close enough to read what it says, you find slogans like "U.S. Out of Nebraska," "Stop U.S. Imperialism in Afghanistan," and "Victory to the Nebraskan People's Movement." "Well, if people don't have a sense of humor, that's their problem," Schneider says.

But Don is not alone in his practical joking. For whatever reason—maybe it's because they can *see* what they're studying, because what they talk about day in and day out has some perceptible *reality* to it—the astronomers seem to be the happy bunch at the Institute for Advanced Study. The mathematicians, with their heads in the World of the Forms all the time, they can be rather grim fellows. And as for the particle physicists, well . . . they've been known to imagine that they're the lords of creation, what with their stupendous particle accelerators, their new-wave superstring theory that even only a few of them can possibly understand. The astrophysicists, though, they don't take themselves quite so seriously. It was only last year's April Fool's Day that Kavan Ratnatunga had a bit of fun with the computer system.

Ratnatunga, a short, swarthy fellow with a mustache and a wheedling smile, is a whiz with the computer, the VAX 11/780 down in the basement. A year ago, on the night before April Fool's, he and Sterl Phinney—who's since gone on to a Caltech professorship—spent a few hours entering a bogus program into the computer. They rigged it up so that when you sat down in the morning and logged onto the system you got a message on the screen that said "Good morning, Bob" (or whatever your name was; they fixed it so that on-screen messages would be personalized according to your log-on code), followed by:

> I am Hal 9000 from the year 2001 A.D. I have come back from orbit around Jupiter to take over all computer systems on earth in order to stop the Star Wars project.
>
> Can I help you?

The befuddled user would then try to get out of the program by typing in the standard exit command—Control Z—but that would only get him the further message:

> You cannot exit this program, Bob.

Bob would then of course hit Control Z again:

> If you try that again, Bob, I will start trashing your files.

Another Control Z . . . and poor Bob would read the truly horrifying statement:

> Now deleting Bob's file #1.

It would be obvious by this time, of course, that some practical joker had gotten into the system and put this dimwit program in there—just a harmless bit of funning—but knowing that by itself didn't help you get out of it. Sooner or later, though, people tried typing in "April Fool," and then—Thank God!—the computer was back to normal. But everyone loved it. The program was a big hit and remains in the system to this day, a monument to Kavan Ratnatunga's whimsy.

Don Schneider's story at the luncheon, by contrast, was not all that implausible. A couple of years earlier Ed Turner, of Princeton, had reported at one of the Institute's Tuesday lunches that he and colleagues at Caltech had discovered a new quasar and an associated gravitational lens. Turner had just returned from a trip to Mt. Palomar, where the observations were taken, and on the morning of the luncheon he had gotten a call from his friends out there who gave him the latest data. The quasar was extremely faint, and its redshift was measured as 3.273, not a record-breaker, but a very distant object nonetheless. At lunch Turner passed around optical pictures of the object taken just two nights before. He brought them back from Palomar himself and had them duplicated in the nick of time for the working lunch. They were all the real thing, no fooling there. That Don Schneider had been able to put one over on the Princeton-area astros merely underlined the fact that these people are by now accustomed to being served all sorts of semi-miraculous news morsels along with Franz Moehn's culinary delights.

A week after the Don Schneider affair, the Princeton astros are convened again for lunch. John Bahcall is up there at the head table, this time together with Charles Alcock, who's visiting from M.I.T., and with Jerry Ostriker, head of the astrophysical sciences department at Princeton. Bahcall gives people five or ten minutes of eating time, and then he bangs the glass for quiet. Piet Hut has an important announcement about a fellow Institute member, Jeremy Goodman, who unfortunately cannot be present: the night before, Goodman's wife gave birth to a baby daughter. *De rigueur*, Hut gives the baby's weight.

"Do you know the dimensions?" John Bahcall asks . . .

Getting down to business now, Bahcall asks Alcock, who's in the middle of his Hungarian veal goulash, to tell everyone what he's learned since the last time he was here. Alcock had spent four years at the Institute as a post-doc, from 1977 to 1981.

While everyone else munches on expectantly, Alcock, speaking with the barest trace of a New Zealand accent, talks about the heavy elements on the surfaces of white dwarfs. White dwarfs are stars about 1 percent the diameter of the sun, and have very high densities, about a million times that of water. Observations of these stars show that floating around on the surfaces of some of them are heavy elements—elements heavier than helium. This is a puzzle because the only apparent source of such elements is the surrounding interstellar medium—the gas between the stars. The problem is, though, that the interstellar medium is mostly hydrogen, the lightest element there is. The heavy elements on these white dwarfs, therefore, can't be coming from the interstellar medium. On the other hand there's nothing else out there for the heavy elements to be coming from, so . . . where *are* they coming from?

Alcock takes a sip of water and gives the audience a moment to mull this one over, a nice dramatic pause there. Finally, "You have to get these heavy elements from *somewhere*," he says, "but from a source that won't add any hydrogen, because there's no hydrogen in the heavy elements. Now the fact is that the asteroids of the solar system have essentially no hydrogen in them, but they contain a lot of the sort of elements that are seen in these stars."

Alcock, in other words, is proposing nothing less than a theory of *extrasolar asteroids*. Astronomers have for a long time entertained the thought of extrasolar planets, but this idea of asteroids circling other stars is something new. According to Alcock, though, the idea is perfectly reasonable. There are asteroids in our own solar system, so why not elsewhere?

"The solar system's asteroids break up and create a lot of dust," Alcock says, "and most of it falls into the sun. So this is a very plausible source of the heavy elements on the white dwarfs."

There are some nods of agreement from around the audience . . . but someone has a question: "Will these asteroids survive the transition from a normal star to a white dwarf?" Before a star becomes a white dwarf, it goes through an intermediate stage of red giantism, when the solar furnace swells up to many times its normal size. Many millions of years hence, when our own sun becomes a red giant, it will balloon outward and engulf the inner planets—probably including earth, which will wind up as a large ball of burnt rubble orbiting inside the sun. So the question is whether an extrasolar asteroid belt would survive the burning red giant.

"Well, in our own system the asteroids probably would not survive," Alcock says. "But in another system, the asteroids might be twice as far away from the center of the star, in which case they would be quite secure."

Lyman Spitzer, the Princeton University astronomer, raises his hand and states that some interstellar clouds have unusual concentrations of elements. Couldn't these interstellar clouds account for the heavy elements on the white dwarfs?

"The answer is that I haven't looked into that in depth," Alcock says. "But the problem with the interstellar clouds is that they're hydrogen rich—so you have the hydrogen problem cropping up again. On some of these white dwarfs there's just *no* hydrogen to be observed at all. So the heavy elements have got to be coming from somewhere else."

The audience is intrigued—quite fascinated, in fact—and at this point, just the right psychological moment, Bahcall thanks Alcock for his presentation, and turns to James Binney.

Binney, here on a visit from Oxford, is an old hand at this stuff: he's been coming to the Institute on and off for a good ten years. In a crisp British accent he describes his latest work on computer simulations of galaxies.

Computer simulations—this is new-wave astrophysics. Almost unheard of twenty years ago, astronomers now model virtually everything on computers, including the birth and death of stars, the evolution of galaxies, the long-term future of our own solar system. The problem with simulations of very large systems—such as galaxies—is that computer models are often dynamically unstable. The model will hold together on the computer screen for a matter of seconds or minutes before it either collapses in a violent death spasm or flies apart like batter off an eggbeater. The challenge is to construct a model that will not only be stable, but will also mimic the observed evolution of star clusters, galaxies, or whatever. Binney has been working on the dynamics of star clusters, groups of some hundreds of thousands of stars held together by their mutual gravitation. Some of these clusters are observed to change shape with the passage of time. They start out more or less as spheres then deform into prolate shapes—into the shape of a cigar, for example.

"There's been work done here at the Institute and at Berkeley," Binney says, "discovering what the range of models is which is unstable in this way, but those calculations have mostly been extremely expensive n-body problems. However, one of my post-docs and I at Oxford have just finished such a calculation where we showed how we could use a rather more elaborate theoretical apparatus to make very much simpler—in practice—simpler, cheaper computations that give actually more accurate results."

When two celestial bodies interact with each other gravitationally—the earth and the moon, for example—the problem of figuring out their future positions is called a two-body problem and is easy to solve. The relevant equations were stated by Newton, and college students work with them routinely in freshman astronomy courses. When a third mass enters the picture—the earth, the moon, and the sun, for example—the complexity of the problem goes up a big notch, and the more bodies you add, the more you increase the difficulty of the problem. Before computers came onto the scene, solving an n-body problem for more than a handful of masses was just about hopeless. Computers, though, can work on thousands of bodies at a time. "You simply program a computer with the initial positions and velocities of quite a large number of stars," Binney says, "like 10,000 for example, and just integrate the equations of motion and allow the system to evolve forward, and you see what happens."

But even on the best computers—and the best ones are expensive to operate and hard to gain access to—a big n-body calculation takes a lot of time and money. Binney, though, has figured out a new way to get better results at a fraction of the usual cost. He's taken an idea from quantum mechanics and applied it to astrophysics.

"The idea is that you express the configuration of the stars not in terms of naive coordinates like positions and velocities, but in terms of coordinates called *actions*," Binney says. "The long and short of it is that you can figure out how the orbits are going to evolve without following the evolution of all the individual components."

Joshua Barnes, Jeremy Goodman, and Piet Hut—all Institute astrophysicists—published a paper in *Astrophysical Journal* early in 1986 calculating the dynamics of star clusters. Here it is only a few months later, and Binney has duplicated those same results at about one five-hundredth the cost. This is a technical advance, not a theoretical one, but Hut—who's in the audience—is very interested. There's no time for him to get deep into it now, though: there are two other speakers still to be heard from, and soon Jerry Ostriker has the floor.

The routine continues until 1:30, when the lunch ends.

During a typical year at the Institute, there will be about two dozen astrophysicists in residence at building E. To make sure that everyone stays abreast of everyone else's work, the astrophysicists have evolved the coffee hour.

It's a bit after 10 o'clock, and eight young astrophysicists are sitting around a conference table in building E's library. The library is a large white room with a gently vaulted ceiling and bookshelves along the walls filled with tools of the trade, which is to say bound issues of *Nature, Astrophysical Journal, Acta Astronomica, Astrophysics and Space Science*, and so on. *Astrophysical Journal—"Ap. J.,"* as they call it—is *the* place to get published if you're a theoretical astronomer. It's the astrophysicist's journal of record, and it comes in three varieties: the standard *Journal*, a *Supplementary Volume*, and a separate issue of *Letters*, for those special late-breaking discoveries that just can't wait. Without exception, everyone in the room has had one or more articles published in *Ap. J.* The other tool of the trade is the computer, and so spread out on one of the two conference tables are the latest issues of *Byte* ("The Small Systems Journal"), *Computerworld*, and *Tugboat* ("The TEX Users Group Newsletter"), a publication devoted to computer typography.

The Institute astros are sprawled out in upholstered chairs around a big conference table, drinking coffee. Moti Milgrom is there, and James Binney, Kavan Ratnatunga, Herbert Rood, and some others. Every now and then one of them will go back to the coffeemaker at one end of the room and get another cupful.

Item one for discussion is the state of the computer downstairs, the DEC VAX 11/780 two floors below in the basement. The VAX crashed early this morning because the air conditioner down there—which exists solely for the benefit of the computer, not for the people—was not lofting out enough of the required cooling breezes, and so the system shut itself down. For most of the people in the room, this is very bad news: it means that their work will just about come to a halt. They won't be able to write their programs, run their simulations, edit their articles, receive their computer mail, and so on. This is a major disaster, and John Bahcall knows it—he uses the computer himself—and so he leaves the room to make some calls: he's got to get the thing up and running again. Meanwhile, Piet Hut, the newest faculty member in astrophysics—he's one of the original Death Star theorists—he and James Binney are getting into an argument about programming languages. Hut thinks that too many languages are created with the machine in mind, rather than human beings.

"The first thing about any program is that it should be *human* readable," Hut says. "The second thing is that it should be machine readable."

Hut is dressed in brown corduroys and brown-and-tan-striped shirt. It's a cold day, but he's wearing sandals. He's got long salt-and-pepper hair and a beard, and looks something like a medieval sorcerer. Hut favors LISP, the language of artificial intelligence theorists, but Binney doesn't like all the parentheses you have to use in LISP. "You end up having to *count* the bloody things," Binney says.

"Well, if you have a text editor," Hut says in his heavy Dutch accent, "it will count the brackets for you, and so there's no problem at all. The thing about LISP is that it forces you to think logically, because it was designed logically. The language invites you to break up problems into as small units as possible. It's infinitely more easy to figure out a LISP program than a program in C, for example. An hour after I learned LISP I was programming in it."

The others aren't impressed. Someone raises the problem of storage allocation. With LISP you have to tell the machine in advance how much memory you're going to need, but how do you know this before you actually run the problem?

"LISP is so damn slow," someone says.

"There are so many dialects of LISP," someone else chimes in.

"There are lots of dialects of many computer languages," Hut says. "LISP is the second oldest computer language, after FORTRAN. LISP is the language that's closest to the way physicists actually think. It has less concessions to the machine."

After a while John Bahcall returns with a relieved look on his face: the VAX will be back in action shortly, he says. Bahcall had heard the tail end of the LISP discussion walking in. "The problem with LISP," he says, "is that no science students learn LISP. Since 1965 people in computer science were against FORTRAN on the grounds that it will soon die out, but it's still the language of choice in science."

"In twenty years no one will be programming in anything but LISP," Hut says.

"Sure," says someone else. "All you need in LISP is seven words . . . and ten thousand parentheses." Everyone laughs. "Discussing computer languages is like discussing politics," Hut says. "People become vicious!"

At this point Jeremy Goodman walks in. "Ah, the father!" Bahcall says. (It was Goodman's wife whose baby was announced at the luncheon the day before.) Everyone gives him a round of applause, and Goodman smiles and blushes. Hearing the uproar, some other people come into the library to see what all the commotion's about, but it's only a baby, not a new galaxy or . . . something *interesting,* and so most of them leave again, but some others remain. They discuss computer modeling for a while, but a little later the coffee hour is over and it's time for Jeremy Goodman's seminar.

Goodman's a junior member of the Institute, only twenty-nine, but he leads a bi-weekly seminar on recent astrophysics for anyone who's interested. He makes regular reading assignments—just like back in grad school—but that's where the similarity ends. The people here aren't in this for credit. They just want to *know;* they want to find out *what's happening out there* in the vast celestial blackness.

Goodman closes the library door—a sign on the outside of it reads "Seminar in progress"—and returns to his seat. There are five people left at the table around him: Piet Hut, of the Institute; Cedric Lacy, a young Princeton University post-doc; Andrew Hamilton, a tall, blond fellow also doing post-doctoral work at Princeton; Bruce Draine, a Princeton faculty member who had been an Institute member back in 1979-80; and Anthony Stark, from Bell Labs.

Bell Laboratories, a few miles away in Holmdel, New Jersey, had given birth to radio astronomy back in the 1930s when Karl Jansky noticed some radio static in the Bell System's transatlantic shortwave-radio-telephone apparatus. The static turned out to be from the center of the Milky Way galaxy. In 1965, also at Bell Labs, Arno Penzias and Robert Wilson discovered the universe's microwave background radiation, the 3° above-absolute-zero remnant of the Big Bang. Tony Stark is one of Bell's latter-day astronomers, come here to puzzle over the structure and evolution of molecular clouds.

Everyone on hand knows what the issues are. Molecular clouds—as the name suggests—are clouds of molecules, mostly of hydrogen, deep in interstellar space. It's known, or at least it's thought on very good observational evidence, that the galaxy's stars are born when the denser clumps of such molecules collapse under the weight of their own gravity. The problem is that identifiable molecular clouds don't seem to be collapsing as fast as they ought to be according to theory. If you look at individual complexes of molecular clouds, you'd expect one of them to coalesce and give birth to a star in about a million years or so, on the average. That's about how long it's supposed to take for the particles within these clumps to feel their mutual gravitational attraction and to start falling inward, ultimately to form a star. Now the total amount of matter in these molecular clouds is roughly on the order of 2 billion solar masses; in other words there's enough free-floating stellar stuff out there to form 2 billion suns. But when you perform a simple calculation—that is, when you divide 2 billion solar masses by 1 million years, which is the time it should take for any given molecular cloud to collapse—you find that stars ought to be forming at the rate of some 2,000 stars per year. The crux of the matter, though, is that observations disclose a stellar birth rate of only a paltry three or four stars per year. That's a factor of a thousand discrepancy between the rate of observed starbirth and the rate that they ought to be forming, on the best theories of how and when molecular clouds give birth to new stars. Theoretical astrophysics, in other words, has run up against the blank wall of empirical observations. Something's wrong somewhere, but where?

At the seminar today, the astros are studying a special case of molecular clouds, those in the constellation Orion. Anthony Stark and his colleagues at Bell Labs are observational astronomers, and they've been "mapping the hell out of Orion," he says. So Stark explains what they've found in that region, a molecular density of such and so, but he acknowledges that much of the cloud is inaccessible to observation.

Someone asks Stark how he can say what the average density of the cloud is if a lot of it is hidden. Someone else wishes they all had a picture of Orion's molecular clouds in front of them, so that they could see what it is they're talking about, and so Goodman gets up and goes to the bookshelves containing back issues of *Ap. J.* He pulls a few volumes down and leafs through them and finally locates the one he wants. After a while he comes back to the conference table with the book open to a map of the molecular clouds in Orion.

Stark recognizes the map right away. "Oh, yeah, it's that old paper by Kutner." (To be exact, it's "The Molecular Complexes in Orion," by M.L. Kutner et al., published in *Astrophysical Journal* in 1977.)

Everyone gets up to look at the molecular cloud map.

"This picture is largely a fiction based on very sparse data," Stark says. He's entirely serious, but everyone laughs at this. Stark goes on to explain how it's possible, with reasonable accuracy, to figure out the general density of a molecular cloud without having observed very much of it.

Listening to this, Andrew Hamilton raises an eyebrow and smiles, as if he's a bit skeptical about it all.

At this point, with everyone huddled over the data, another seminar member walks in, very late. This is Philip Solomon, of the State University of New York at Stony Brook. He's older than the others, in his late forties, balding a little. Solomon was at the Institute during 1973–74 and he's back now for the spring semester. He's an observational astronomer with a heavy interest in theory, and everyone turns to him now as if he's going to be their salvation. Goodman explains the problem at hand, and Solomon immediately states his opinion.

"Well, the collapse rates are not the same for every cloud," he says. He goes to the blackboard and draws a picture of a molecular cloud and, below it, a histogram showing the molecular densities in different regions. Piet Hut, who's been quiet for a while, doesn't like this picture at all and goes to the blackboard to draw a different one.

"Oh no, it's certainly not like *that*," Solomon says, shaking his head. They stand there for a while, each correcting the other's picture, and then they both return to the conference table.

"Well, let's turn to the paper," Jeremy Goodman says.

They all have a paper in front of them—it was part of their "homework" for today—about the molecular cloud problem, but most of them acknowledge that they didn't find it too helpful. "All this basic freshman stuff about gravity in here, and it gets published in *Ap. J.*, I couldn't believe it," says Bruce Draine.

"Well, shall we move on to the next paper?" Piet Hut asks. But they don't. They keep on with the first one, trying to make sense out of one of the charts.

The scientists argue for another half hour, trying to figure out why it is that these molecular clouds aren't collapsing as they should. They talk about "column densities," "crossing times," "core collapse rates," and so on, but they don't seem to be making any progress. Now and again someone pauses to ask, "But do we really *know* that?," underlining the fact that all of them are trying to theorize on the basis of incomplete data, unproven hypotheses, and untested assumptions. By the time the seminar ends, a little after noon, the only thing they can agree upon is that maybe the molecular clouds aren't as dense as everyone had thought to begin with. But the original mystery, the gulf between observed stellar birth rates and the rate predicted by theory, this mystery remains.

Astrophysics didn't become a big subject at the Institute until after Oppenheimer stepped down as director in 1966. Oppie concentrated on particle physics and brought in some of the best brains in the business, which is quite understandable. Particle physics, after all, was where the main action was. Quantum theory had produced major upheavals, and the atom bomb, well . . . that was in a class by itself as far as crucial scientific experiments went. Cyclotrons were springing up all over the globe and experimentalists were finding new particles one after the other, and coming up with new theories to explain the particles. The whole field was in a constant state of pandemonium, and it was a battle for the theorists to keep up with the results the experimentalists were handing them.

There were no such revolutions occurring in astronomy, however, which seemed to be progressing about as it always had, by process of slow and stately accretion. And there was a good explanation for this. For one thing, while astronomy is often regarded as a branch of physics, it's not an experimental science in the same sense that the rest of physics is. In physics proper, theories are testable by laboratory experiments. Some of these experiments may require huge accelerators, but the accelerators have the virtue of at least settling the issue once the experiment is run. For the most part, however, the same thing doesn't hold for astrophysics, because there's just no way to get out there and handle the objects themselves—the stars and galaxies—in order to test theories directly. So observational astronomers have to be content with sitting at their telescopes and passively gazing off into the nether reaches. For this reason astronomy has more in common with geology—where sheer looking also takes a back seat to experimenting—than it does with sciences that are testable in the usual sense.

A second reason for the slow advance of astronomy is that its data-gathering instruments—optical telescopes—hadn't changed much in 200 years. At the end of World War II, the world's largest telescope was the 100-inch reflector at Mount Wilson Observatory outside of Los Angeles. But a century and a half earlier, in 1789, astronomer William Herschel was already looking through a reflector fully half that size, the 48-inch instrument that he built in England. So astronomy had a slowly growing body of data, lots of theories, and no way to test most of them.

After World War II, however, a revolution occurred in astronomy that was comparable to the one produced by accelerators in particle physics. This was the birth of radio astronomy. All the telescopes from Galileo onward had gathered data in the form of visible light rays. Celestial bodies, however, emit a lot more than visible light: they broadcast streams of radio waves, X-rays, gamma rays, and infrared and ultraviolet rays as well. Previous to radio astronomy, observers were taking in only a small portion of the incoming data, while the rest of it went flying past them like so much chaff on the wind. It was as if astronomers were colorblind, seeing only shades of gray while a fully technicolor universe blossomed above their heads. When radio receivers started springing up like mushrooms in the 1950s and '60s, scientists got a whole new line on the universe.

The radio sky is no mere carbon copy of the visible sky. Objects that are optically bright—like the moon and the planets—are silent at radio wavelengths, while those that are invisible to the eye—such as the huge clouds of matter on either side of a radio galaxy—are bright in the radio spectrum. With the development of X-ray and gamma-ray telescopes, infrared sensors and neutrino observatories, astronomers had grown new senses.

Later, in the 1970s, there was another revolution: the computer gave astrophysicists a measure of experimental control over their subject matter. After hundreds of years of being able to do nothing but stare at things from light years away, astronomers were now manipulating miniature versions of celestial entitites ranging from orbiting asteroids to individual stars, clusters of stars, galaxies, clusters of galaxies, even—quite immodestly—the universe as a whole. It was a whole new way of poetizing the phenomena, and a new expression of the scientist's hubris. There's hubris enough in thinking that a small group of individuals, on one planet in a vast system of stars, can know the shape and structure of immense objects at extreme distances, but there's another rash dose of presumptuousness in thinking that science can reduce an entire galaxy, indeed the whole physical universe, to the rush of electrons through a few silicon chips.

With the universe opened up by radio astronomy, and with computers giving theorists working models of the things in themselves, the Institute for Advanced Study began taking astrophysical science seriously, and in 1971 hired John Bahcall as professor. When he became executive officer of the School of Natural Sciences, he was not at all bashful about packing the place with astronomers. These days, when Institute administrators have to scramble to find office space, or rooms in the housing project, for visiting members, they blame it on John Bahcall, who brings in astronomers by the dozen. The Institute's astrophysicists, coupled with the university astronomers next door, have made Princeton a world center for astrophysics. "The Institute is the best place in the world for stellar dynamics at the moment," says Douglas Heggie, visiting from the University of Edinburgh. "There's a lot of very interesting work being done here, and every time I come for a visit I hear something which surprises me." James Binney agrees. "In my particular area, Princeton is the best place in the world by far. I've tried Cambridge and I've tried California, and I don't think they're even in the running."

Institute astrophysicists study everything from the sun to the most distant quasars, but many of them concentrate on stars and star clusters, and objects farther out—"extragalactic," as they say. You don't find too many people in building E doing planetary science—studying the planets and satellites of our own solar system—although of course there are a few exceptions, such as Scott Tremaine. Tremaine had investigated the structure of Saturn's rings and predicted that in order to account for their perceived structure there had to be satellites circling the planet over and above those that were already known. Sure enough, when the Voyager I spacecraft encountered Saturn in 1980, the "sheepdog" satellites were in fact right there, orbiting exactly where Tremaine said they'd be.

And then there's Piet Hut, of Nemesis—"the Death Star"—fame. It was Hut who, in 1984, along with Marc Davis and Richard Muller, proposed that there may be another star in our own solar system, an unseen companion of the sun. Davis, Hut, and Muller analyzed the apparent periodicity of mass extinctions on earth, which seemed to occur again and again every 26 million years or so, and decided that this pattern could be explained by an unseen companion star on an eccentric 26-million-year orbit around the sun. The star would careen out to a distance of some 3 million light years, then come back in for a close encounter with the sun before heading back out of the solar system. On its in-and-out travels, the Death Star's gravitational field would dislodge some comets from their orbits far out in the Oort cloud, causing them to crash to earth, where they'd produce a nuclear-winter-type period of cold and darkness that would kill the dinosaurs and other species in a giant wave of mass extinction.

"If and when the companion [star] is found," wrote Hut and his colleagues in *Nature*, "we suggest it be named Nemesis, after the Greek goddess who relentlessly persecutes the excessively rich, proud, and powerful. We worry that if the companion is not found, this paper will be our nemesis."

The companion star has not been found, but in 1985 Hut was made an Institute professor, not only for the Nemesis proposal, but for his work on the dynamics of star clusters and the three-body problem.

In astrophysics, theory and observation are closely intertwined, and many of the members have actually spent some hours in front of the telescope. There are no telescopes at the Institute, so taking observations means leaving the premises. Often enough, new observations will overturn old theories, something that astronomers find they have to get used to.

"Astrophysics is a very young science," James Binney says. "The information we have per unit of junk out there is very limited, so by and large we don't have very many connections between the model and the reality. Few of us would ever turn over and drop dead with amazement, therefore, if we came across something which didn't fit the pattern. In astrophysics, one thing wrong and we begin to feel nervous. Two things wrong and we begin to feel that, well . . . maybe we better have a complete fresh think about the whole problem."

Tuesday has rolled around again, the day when the astros themselves seem to be in orbit around Princeton, with the university people coming to the Institute for the working lunch, the Institute people going over to the university for the afternoon lecture. Although the same outside speaker will be featured at both events, there must be no repetition in the two talks because the audiences will be exactly the same. But today will be a special day for the Princeton astros, even for a Tuesday: Margaret Geller is coming down from the Harvard-Smithsonian Center for Astrophysics, in Cambridge, Massachusetts, to tell them the real truth about all these . . . *bubbles* they've been hearing about. Geller and two colleagues at the Center for Astrophysics, John Huchra and Valérie de Lapparent, have collectively shown that the universe's galaxies appear to be distributed, not more or less randomly as had previously been thought, but on the surfaces of cosmic bubbles.

The astros have already heard about this, of course, through the standard communications grapevine. Geller had already presented their new findings a few months ago—back in January, it was—at the American Astronomical Society meetings in Houston. Before that, astrophysicists on the preprint line received an advance copy (a "preprint") of their joint article "A Slice of the Universe," before it appeared in *Astrophysical Journal (Letters)*. And if they somehow happened to miss all that, they would still have heard about this bubble theory from the media, for it appeared everywhere from *Time* magazine ("Bubbles in the Universe," the headline said), to out-of-the-way dailies like the *Green Valley News-Times* of Tucson, Arizona ("Lawrence

Welk Universe Theory"). Still, not everyone is absolutely convinced that the universe has a large-scale cellular structure, and so Geller is coming to Princeton to present the message in person. Word has it that she's fully armed . . . with slides, view-graphs, even a movie.

At the working lunch Margaret Geller is up there at the head table with John Bahcall, Jerry Ostriker, Ed Turner, and Jim Gunn. Geller, in her late thirties, is trim in a businesslike black suit. Bohdan Paczynski, the organizer of the Princeton University lecture series, is at one of the side tables, and John Bahcall asks him to announce the afternoon talk at Peyton Hall.

"Yes, very good," says Paczynski, in his Polish accent. "It will be at 2:30. The lecture will be by our guest Margaret Geller, and is entitled 'Bubble, Bubble Toil and Trouble.' I understand that a movie will also be shown."

"It's rated PG, is that right?" Bahcall asks.

Geller, who's just put a forkful of lettuce into her mouth, nods her head. "PG is right," she finally manages to say. "Bring the children."

Bahcall asks Philip Solomon to report on his recent molecular cloud observations. "This is real data," Bahcall says, "not computer simulations."

Solomon passes around two sheets full of data and starts in with his talk, which summarizes much of what he had said in Jeremy Goodman's seminar the day before. He speaks for about ten minutes, then fields questions.

Jerry Ostriker and Lyman Spitzer, both Princeton faculty, get into a polite argument about something Solomon's said and Bahcall tries to break this up by banging on his water glass. After a while he's successful. The audience is impatient to hear from Geller, even though this will not be the bubble lecture, and finally Bahcall turns to her.

"Everyone's looking forward to your Shakespearean lecture this afternoon," Bahcall says.

"That's 'Much Ado About Nothing'?" Jerry Ostriker chimes in.

Geller's enjoying all this immensely. She'd been a grad student at Princeton, where she got her Ph.D. in 1975, back when Institute lunches were off limits to students. Returning now in a different role, this quip-and-banter stuff is right up her alley.

"You know, my colleague John Huchra has been observing redshifts near the wide gravitational lens," she begins, "and he found this cluster of quasars with a redshift of 4.1. . . ."

Aha! News of the Institute's fake 4.1 quasar has gotten all the way north to Cambridge, Massachusetts. (Later on, Ed Turner says that it also got as far south as Charlottesville, Virginia. He'd gotten a call from astronomers at the university down there who wanted more details.) Finally, though, Geller launches into her talk.

She and her colleagues plan to explore the consequences of their bubble findings, Geller says. The main thing they want to do is to confirm whether the picture painted by redshift observations bears a good relationship to what's really out there in three-dimensional space. They'd found bubbles in the universe by gauging the distances of galaxies according to their redshifts, meaning the degree to which the light coming from the galaxies is shifted toward the red end of the spectrum.

"Redshift is a measure of three things," she explains. "It's a measure of the cosmological expansion, it's a measure of velocities in bound systems of galaxies, and it's a measure of large-scale flows. All those are included in the redshift. Now you want to know the relationship between the structure in redshift space and the structure in real space, and we don't exactly know that, so we need to make some estimate of relative distances by other means."

Geller describes the planned observations in technical detail, and then there's the usual question-and-answer session. Finally, though, time's running out. John Bahcall turns to Bohdan Paczynski.

"Bohdan, why don't you say something amusing before we go?"

Paczynski is not at a loss. He's been thinking about the problem of gamma ray bursts, and he's got a new suggestion.

Well, this is not surprising. Gamma ray bursts are short episodes, comparatively rare and lasting only a few seconds, with only four or five of them observed per year. Most everything about them is completely uncertain, including what causes them, whether they originate from far away or close to earth, and astronomers have had a field day coming up with competing, often bizarre, explanations.

"As you know," Paczynski begins, sounding a little like Bela Lugosi, "these things have been attributed to just about everything up to and including signals from extraterrestrials in distress."

One problem in identifying them, he says, is that there's still no way to determine where they come from, whether they're from the Oort cloud region at the outer edge of the solar system, or from "cosmological distances"—meaning from outside the galaxy.

Bahcall interrupts: "Even for a working luncheon, Bohdan, that's a rather high degree of uncertainty."

Paczynski agrees, but he comes out with his new idea. Perhaps the bursts are produced by comets plunging into neutron stars. That would produce a sudden pulse of gamma rays that would be observed from earth. It would explain all the data perfectly well.

People start turning this one over in their heads.

Philip Solomon breaks the silence. "Now the only remaining question for this audience," he says, "is whether the comets hitting the neutron stars are killing the dinosaurs."

Geller's afternoon lecture is being given in the auditorium of Peyton Hall, home of Princeton University's astrophysical science group. It's a tiered lecture room, with banks of student-type chairs with the little fold-up desks to take notes on. The talk begins at 2:30, but at 2:25 there's not a soul in the lecture hall, except for Paczynski who's fiddling with the film projector up at the back of the room. At 2:27, though, the astros start pouring in and by 2:31 the place is three-quarters full. All the Institute members seem to be in attendance: the two Bahcalls, John and his wife Neta, Tsvi Piran, Jeremy Goodman, Don Schneider, Kavan Ratnatunga, Stefano Casertano, James Binney, and all the rest—everybody, in fact, except for Piet Hut, who's in Japan attending a conference. The Princeton astros are there, of course: Jerry Ostriker, Lyman Spitzer, Ed Turner, Jim Gunn, and Jim Peebles.

Peebles literally wrote the book on what the universe looks like, *The Large-Scale Structure of the Universe*. He finished it during a sabbatical year at the Institute, 1977–78. "When I wrote my book," he says, "there were rumors—hints—of these filamentary or sheet-like structures, but it was borderline stuff, and I was a skeptic. You'll find very little about it in my book."

The Princeton post-docs—Andrew Hamilton and Cedric Lacy—have their notebooks at the ready, and in addition there's a handful of grad students. In all, a total of some seventy-five people have come to hear Margaret Geller reduce the universe to suds.

At the front of the room, Paczynski makes the introduction. "The speaker for today is Margaret Geller, from the Smithsonian Center for Astrophysics. Her talk is entitled 'Bubble, Bubble Toil and Trouble.'"

Geller, who's been sitting in the front row, gets up.

"Charles Alcock invited me to give this talk once at M.I.T.," she begins, "and he asked me for the title of it, and I told him 'Bubble, Bubble Toil and Trouble.' He said, 'Are you sure it's not "*Hubble*, Bubble Toil and Trouble?"'

Enormous laughter. Edwin Hubble, of course, was the early twentieth century astronomer who made possible most of the knowledge we're about to receive. For one thing, he discovered the fact that there's a whole universe beyond what appears in the sky at night; that is to say, he discovered that there are other galaxies besides the Milky Way.

With only a few exceptions, the naked-eye objects in the night sky are the stars of our own Milky Way galaxy. Some starlike objects, however, have a fuzzy and indistinct appearance, as if they were being seen through a fog or veil. They look like luminous smudges instead of pinpricks of light. Astronomers used to call these fuzzy light sources nebulae, and regarded them as clusters of dim stars, or clouds of gas and dust illuminated by the stars shining inside them. At any rate, astronomers were convinced that these nebulae were really just parts of our own galaxy.

In 1924, though, Edwin Hubble, working at the 100-inch telescope at Mount Wilson, resolved individual stars within what was then called the Andromeda nebula, now known as the Andromeda galaxy: the stars were so dim that they had to be very far away. Hubble found out just how far when he discovered Cepheid variables—stars used to measure astronomical distances—among the stars of Andromeda. He calculated that the nebula must be 800,000 light years away, about eight times as far as the remotest star of the Milky Way galaxy. (More recent estimates put the distance at over 2 million light years.) The Andromeda "nebula" was in fact an entire galaxy unto itself.

Hubble pioneered the study of galaxies, and in the process exploded the astronomer's conception of the size and composition of the universe. He classified galaxies into the now-standard scheme of spiral, barred spiral, elliptical, and irregular, and, more important, he discovered the law of redshifts, now called Hubble's law. Using it, he found that the universe is no static, ageless structure, but rather that it is expanding, growing, swelling outward, that the galaxies are racing away from each other like spots on an inflating balloon. Hubble's law—which states that galaxies are speeding away at rates proportional to their distances, the farther the galaxy the faster its recession—is now a staple of astrophysics. The new picture of the universe that Margaret Geller is about to present is largely based on the techniques first used by Hubble.

Geller waits for the laughter to die down, then launches into her main theme. She's going to be concerned with the large-scale structure of the universe. What is it like? How does it form? How does it evolve? The traditional answers, developed during the galaxy mapping programs of the mid-twentieth century, are based on the doctrine, which had become an article of faith among astronomers, that, although galaxies are clumped together in clusters and groups of clusters, the groupings are distributed more or less randomly throughout space, with no apparent regularities.

Geller projects a graph on the screen. To the untrained eye it looks like the Milky Way as it would appear on a very dark, clear night: a dense strip of dots running across the visual field. It's not the Milky Way, though, for in this case each dot (actually, each one is a small +) is not a star but a galaxy, and there are some 19,000 of them spread out across the screen. Except for a threadlike concentration of them along an S-shaped curve, the galaxies look evenly distributed, just as the doctrine of homogeneity requires. No bubbles show up anywhere, and with good reason: the picture on the screen is a plot of galactic distribution in two dimensions, in which any bubbles that exist would not be readily apparent (see Figure 6).

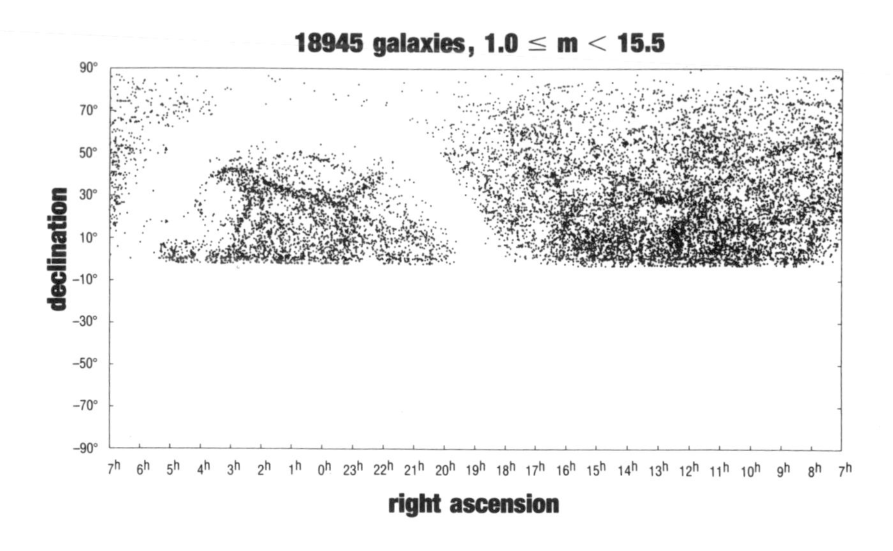

Figure 6 Two-dimensional plot of 18,945 galaxies

The survey that Geller did with Huchra and de Lapparent took a completely different approach to the business of mapping galaxies. Instead of plotting galactic positions in just two dimensions—how far up and how far over—they had added a third dimension: how far back from the observer. This third element turned out to make all the difference. When you plot a three-dimensional projection of the results, the universe's cellular structure stands out in bold relief.

Geller projects another chart on the screen. Their new observations, she says, were taken with the 60-inch reflecting telescope at Mount Hopkins, in Arizona. This new chart is wedge-shaped, like a large slice of pizza. In fact, it's a plot of galactic positions in a three-dimensional "slice of the universe," a 6-degree-thick strip of space extending across a third of the sky and penetrating some 450 million light years into the cosmos.

Some 1,100 galaxies appear on the screen (see Figure 7). They're far from evenly distributed. Rather, they're arranged around the surfaces of what appear to be large holes or bubbles, gigantic voids out there in space.

"When we saw this," Margaret Geller says, "it was clear that we were receiving a message from the universe. But the question was, what does it say?"

The projector hums and the audience stares for a while at the holes in space.

"There are lots of these voids," Geller says. "We found them everywhere. The obvious question was, why weren't they seen before?"

She gives two answers. One, previous surveys didn't see deeply enough into the universe; two, they didn't measure enough galaxies for their bubble-like arrangement to become fully apparent. The eerie thing is, though, that the barest hint of these bubbles was already there in earlier data. Geller takes another transparency and projects it on top of the one showing the bubbles. This new chart, whose galaxies are colored to show up in green, is part of an older survey, done by Marc Davis and others. The green dots are less dense than the profusion of galaxies on the 3-D chart, but there's no doubt that the green dots trace out the ghostly outlines of cosmic bubbles. The evidence was there, but no one really *saw* it until now.

The audience is clearly hooked—the pictures are just that striking—but the best is yet to come. The Center for Astrophysics is supported in part by Congress, and for one of the Smithsonian's budget reviews, Geller and her associates made a movie that presented their findings so graphically that anyone—even congressmen—would have to be impressed. She shows

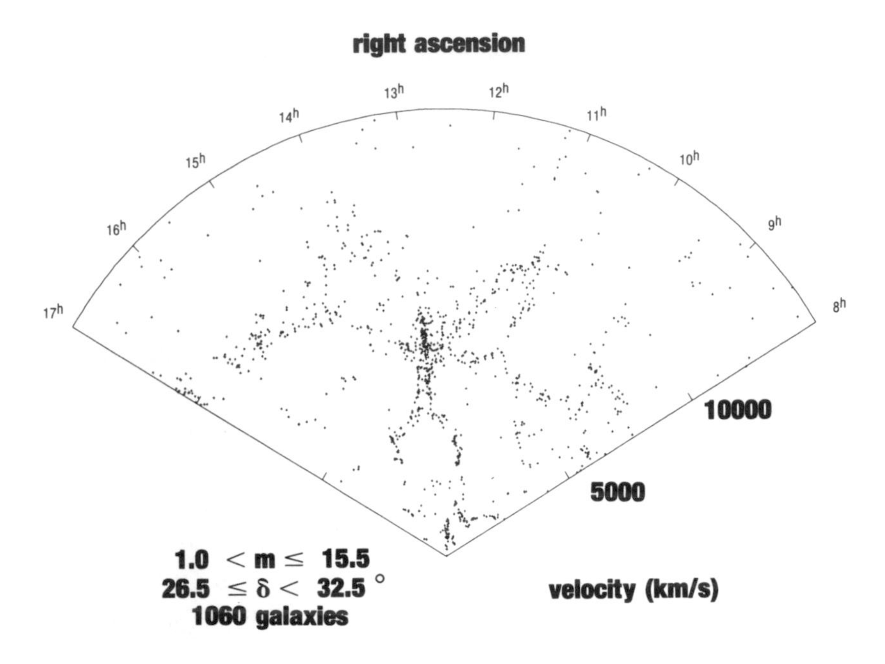

Figure 7 Three-dimensional "slice of the universe"

the movie now. It's a short film, only about a minute long. There's no sound track: the pictures speak for themselves.

The projector comes alive and the credits roll past—"Smithsonian Astrophysical Observatory"—followed by the title: "Bubbles in the Universe." A few people in the audience cheer and whistle at this.

The pie-shaped graph that we've already seen appears on the screen. Now in living color, the galaxies show up as pink dots and seem to float there in a neon-like glow.

The wedge-shaped slice of sky has a depth to it now, for this is a computer display of the data, projected in three dimensions. And then the whole thing, the whole slice of pie, starts to turn, on an axis, and the 3-D aspect of the whole thing becomes utterly undeniable. It's as if we're on the edge of the world and looking inward as the entire universe revolves in front of us. It's a surreal experience, but there are these gaping Swiss-cheese holes out there in the vastness. "The universe is not only queerer than we suppose," J.B.S. Haldane once wrote, "but queerer than we *can* suppose."

Indeed. The slice of the universe completes a full revolution and now the audience is stunned, just about speechless.

The lights come up.

Even Jim Peebles has been converted. "Margaret's survey," he says later, "convinced the last of us skeptics that there's a bubble-like nature to the distribution of galaxies."

End of show. End of another's day's work at the extremes of science.

Chapter 8

Carrying the Fire

The universe that astronomers gaze into is mostly empty space. Indeed that's why astronomers can see so far into it: there's nothing in between blocking the view, and so they can peer through their telescopes and see to the ends of space and time. But inward bound, the view is quite similar. Matter itself, although it has a look of solidity and denseness about it, is also mostly nothingness. This is apparent if you consider that on a good day in Alaska the visibility may be 200 miles or more. You can look through 200 miles of air—through billions of molecules of nitrogen, oxygen, water vapor, and all the rest—and what you *see* is . . . nothing at all. Air is less dense than solid matter, of course, but you can see through solid matter too, as when you look through window glass or gaze into an ice cube or a diamond. How come? Why is it that you can see right through brute things, just as if they weren't there? The answer is that material objects are mostly . . . emptiness.

Matter, as we all know, is composed of molecules, and molecules are made up of atoms, minuscule solar systems whose dense inner nuclei are surrounded by orbiting clouds of electrons. The electrons are extremely distant from the atom's inner core, about 100,000 times as far away from the nucleus as the nucleus itself is wide. Since volume goes as the cube of diameter, the volume of the nucleus—the space that it actually takes up—is less than one trillionth (1/1,000,000,000) the volume of the whole atom. But this is just another way of saying that atoms are essentially blank and barren voids.

Since atoms are mostly empty space, so are the material objects—tables and chairs and everything else—that go to make up the world we live in. If it were possible to remove all the emptiness from an object—squeezing it out like water from a sponge—and to pack its atomic particles together so that they were touching each other like kernels on a corncob, then the world's ordinary macroscopic objects would be reduced to their bare minimums. Most of them would vanish without a trace: a baseball would become an invisible speck, a person would be about the size of a fruit fly, an elephant would be reduced to less than a thimbleful of atomic stuff.

The matter in ordinary physical objects, their component atomic particles, are spread out so thinly that it's as if objects had no greater density than a puff of smoke. Which leads to a great mystery. Why is it that the things of this world look and feel so rigid and rock-solid? You can walk through a puff of smoke with no resistance at all, but when you step on the floor, the floorboards push back and hold you up. And when you lower your hindquarters onto the kitchen chair you don't sail right on through the seat, the floor, and all the way through the earth . . . but in view of all the empty space in there it's rather mysterious why you don't.

And that isn't even the worst of it. If matter is all that empty, why should it take a footfall in order to make the floorboards collapse? Why don't they cave in of their own accord, under their own weight? Why doesn't the whole world just . . . shrivel up and wither away?

What's true for large groups of molecules—such as go to make up the material objects of everyday life—is true for individual atoms as well, and so a person might get to wondering what it is that keeps an atom's electrons and protons apart from each other. Protons and electrons have opposite charges, and opposite charges attract, so . . . why don't *individual atoms* collapse? Why don't *they* puff themselves out of existence? The fact is that it was just this question that led physicists to invent quantum mechanics in the late 1920s. The problem of the atom's stability led Niels Bohr to come up with the idea of quantization of orbits, and Erwin Schrödinger to develop the wave equation. According to quantum mechanics, the reason why individual atoms don't collapse is that electrons have minimum orbits or energy levels below which they just can't fall. Once an atom is in its ground state—its lowest energy level—then it can't go any farther down, and so atoms are automatically stable.

Unfortunately, it wasn't noticed for almost half a century that although this saves the day for individual atoms the problem still remains where large groups of them are concerned. The fact that an *individual* atom's electrons will stay in minimum orbits around the nucleus doesn't by itself guarantee that any given *assemblage* of atoms will remain stable over time. After all, there are still all these yawning empty spaces for other atoms to fall into, and then there are these attractive forces between atoms—Van der Waals forces they're called—pulling them in toward each other. The Van der Waals forces act just like gravity in that, as they draw atoms closer together, the forces between them get stronger and stronger, thereby producing ever more concentrated buildups of atoms, until—*whomp*!—the whole business implodes like a black hole in space.

This, anyhow, was more or less the scenario advanced by two mathematical physicists, Michael Fisher and David Ruelle, who described the problem in their article, "The Stability of Many-Particle Systems," published in the *Journal of Mathematical Physics* in 1966. After considering the case from all angles they decided that they couldn't explain why it was that ordinary, everyday objects wouldn't just go up in smoke. It was quite anomalous. Inert matter just lies there in a kind of steady state, ticking over silently in a metaphysical sort of way, but this ordinary and plain fact of experience was incomprehensible to physics.

In Princeton, theoretical physicist Andrew Lenard started working on the problem while he was at the University's Plasma Physics Laboratory. He thought that there ought to be some way for physicists to understand the stability of matter, to *prove* that physical objects wouldn't vanish into thin air. You ought to be able to apply the laws of electrostatics, he thought, to show that the various electric forces among groups of molecules would add up and mutually cancel each other out in such a way that a material object would persist through time and not fold up like a house of cards. He thought and worked and calculated, but he didn't seem to be getting anywhere. During the academic year 1965/66, Lenard took a year's sabbatical leave from the Plasma Physics Lab, and went to take up a residency at the Institute for Advanced Study. There he'd have all the time in the world to think about the stability of matter.

At the Institute, Lenard began to realize that quantum mechanics would somehow have to enter the picture, and he remembered that Freeman Dyson, the place's star particle physicist, had written something about quantum mechanics and the energies of atomic particles. "So one day I went in to see him and asked him for a reprint of the paper," Andrew Lenard remembers. "He asked me why I needed it, and I told him about the stability of matter problem. Right away he got interested in it himself. He asked me if I minded if he started to think about it too. Of course I said no."

Lenard had come to the right person. The Institute was brimming over with elementary particle physicists of the most academic variety, which is to say that they didn't go out of their way to connect up their theoretical entities with the real world out there. They knew all about the most arcane and exotic elementary particles, they could tell you everything you never wanted to know about the hyperfine splitting of a μ-mesic atom, but to relate their abstract theory to the world of everyday objects, well, this is nothing they had any experience with, and little training for, to say the least. But Freeman Dyson, he was different. He was at home in the everyday world, not just in the universe of pions and kaons. He could think just as happily about designing nuclear reactors or building a spaceship as he could about the recesses of quantum theory. He seemed to be equally conversant with the ways of matter on every level, macro or micro.

Which was much to the good fortune of Andrew Lenard. For a few weeks after Lenard went in to see him, Dyson had figured out why matter wouldn't go up in smoke.

Freeman Dyson is without a doubt the best-known living member of the Institute for Advanced Study. In fact, in these relatively lean years for its particle physics program, with no more Nobel prize winners on the faculty, and no one else putting the place on the scientific map, Dyson has a position comparable to that of Einstein back in the good old days: he gives his aura to the Institute instead of vice versa, and so the Institute is known, in the popular mind at least, as the place where Dyson works. "Oh, so you've been at the Institute?," people say. "Well then, you must know Dyson, right?"

Dyson has always been on the controversial side, perhaps even a shade notorious. Unlike the rest of his Institute brethren, he doesn't remain in any one discipline for very long; rather, his mind ranges all over the place, as if there are too many interesting things in the world to spend very much time on any one of them. Dyson isn't *just* a particle physicist, although it was he who, back in the 1940s, unified the three then-extant theories of quantum electrodynamics (QED). And he isn't *just* an astrophysicist, although he's written papers about neutron stars, pulsars, theoretical galactic dynamics, and so forth. And he isn't *just* a theoretical mathematician, although his degree, from Cambridge University, is in mathematics. He's all these things and much more besides.

Freeman Dyson has probably the most fertile imagination of anyone who's ever been at the Institute, bar nobody. Ideas, schemes, plots, just come pouring out of his head, not only about elementary particles, stars, galaxies, but about . . . *everything*. Consider his scenario for planting trees on comets and riding them around the solar system; or his idea for a reflecting telescope with a rubber mirror (it could be reshaped to compensate for atmospheric distortions); or his proposal for reprogramming a turtle's

DNA so that it would grow diamond-tipped teeth. Fleets of these jewel-choppered tortoises would be sent waddling up and down America's highways to eat up tin cans, bottles, and empty hamburger wrappers. They may be a little crazy, of course, but anyone flat out has to admit that . . . *the man has ideas!*

Dyson has a sense of perspective about this stuff, as well as a sense of humor: he knows it's all fairly nutty. But to Freeman Dyson nuttiness is not a vice but a virtue. "Have you ever been to Cambridge University?" Dyson once asked. "It's full of crazy people—oddballs, loners on the verge of doing something really tough and historic. Why shouldn't they be crazy? Nature is crazy. I would like to see more crazy people here at the Institute."

Freeman Dyson came to the uncrazy and quite mainstream Institute for Advanced Study in 1948, when Oppenheimer was director. He came back again in 1950, and a few years later Oppie made him a full professor. Oppenheimer thought he was hiring a particle physicist, someone, as he said, "with a future of as brilliant promise as any man in theoretical physics." But just a few years after becoming a professor, Dyson went off on a leave of absence. His dreams came calling . . . craziness was beckoning from out of the west . . . and from the stars. And so Dyson went out to San Diego to help build a spaceship. It was a special kind of spaceship, of course. It was to be powered by H-bombs.

Dyson's space travel dreams went back to his childhood days in England where he imagined himself traveling to other planets—really going out there, in the flesh, personally. The H-bomb spaceship, however, was not his own idea. It came from the mind of Stan Ulam, the Polish mathematician who had been at the Institute himself for a semester, back in the '30s. Then he had taken various other jobs, winding up at Los Alamos during the war years, and remaining there ever since. He was, together with John von Neumann and Edward Teller, one of the inventors of the H-bomb.

Around 1955 Ulam and his friend C.J. Everett had written a paper about a spaceship propelled by successive thermonuclear bomb explosions. The craft would blast its way through the heavens by setting off H-bombs behind it and riding the force of their shock waves. It was like putting a firecracker under a tin can and watching it blow sky high, but the idea was taken quite seriously by an Air Force committee, and the Atomic Energy Commission even patented the idea.

Private enterprise got into the act when the General Dynamics Corporation set up a division in La Jolla, California, called General Atomic Laboratories. General Atomic was in the nuclear power business, but in 1957, with Sputnik emitting beeps from "outer space" and people already talking about going to the moon, the company thought it would make good economic sense to look into the nuclear-powered spaceship idea. Ted Taylor, a Los Alamos designer of some handy little thermonuclear devices, joined General Atomic and started thinking about using his bombs to drive spaceships around the solar system. He named the undertaking Project Orion, after the Orion constellation.

Taylor had known Dyson from Cornell—both of them had been students of Hans Bethe—and he gave Dyson a call at the Institute and asked him how he felt about bombing himself into orbit. "It sounded good," Dyson said later. "It didn't frighten me. The immediate reaction of everybody is that it will blow the ship to pieces. I wasn't bothered by that. The thing made sense on a technical level. It sounded like what we'd all been waiting for."

What we'd all been waiting for?

Dyson, anyway, had been waiting for something like this, so he took a year's leave of absence from the Institute and moved to California with his family. This was in 1958. The slogan for Project Orion was "Saturn by 1970."

And it wasn't merely a slogan: they all took their work quite seriously and for a while Dyson thought that he really might get to make the trip. He'd look through the telescope in Ted Taylor's back yard and imagine himself gliding through Saturn's rings, then making a soft landing on the satellite Enceladus. There would be water in abundance on Enceladus, enough to start his own hydroponic vegetable farm.

The spaceship on which he'd travel would be a giant pogo stick mounted on top of a pusher plate. Nuclear explosions would be set off underneath, and as the bombs went off the pusher plate would spring up toward the ship, where the force of the blast would be arrested by an elaborate set of shock absorbers. The upward force would loft the spacecraft up and away and pretty soon you'd be sailing past the moon and Mars, and setting your sights for the asteroid belt.

That, at least, was the idea. To any calm and reasonable person, of course, this scheme is no more and no less than the village idiot's conception of space travel, just one step up in sophistication from Jules

Verne's story of firing men to the moon from out of a big circus cannon. The surprise, though, is that this village-idiot plan would work.

In fact it *did* work, at least in small-scale trial runs. Naturally there were the usual failures, just like in the official U. S. space program run by NASA. The Project Orion tests took place off Point Loma, a high cliff overlooking the Pacific just west of San Diego. Ted Taylor and Freeman Dyson and their crew would go out there on Saturday mornings where they'd watch their Orion test model rise up and . . . get blown to smithereens.

At the beginning even that didn't happen. The first tests used a working model with a one-meter-wide pusher plate. The bomb went off—just an ordinary chemical explosive, not an H-bomb—but the scale model just sat there. The explosions were going off under it a mile a minute—*kaboom!, kaboom!, kaboom!*—but the thing just sat there as if it were made out of stone. "I think we should suspend these tests," Dyson said at the time, "unless we can achieve over one-G acceleration." Heh, heh, heh.

Clearly the rocket had to be made lighter—so that it would at least move, so that it would at least *budge* a little—and so they stripped the thing down to its bare essentials and tried it all again. So they would come out on Saturday mornings and run tests of the new souped-up, lightweight model, which they called "Hot Rod."

At least Hot Rod moved. It would spring up off the launch tower, zoom up into the sky, and then—*kaboom*!—it would explode in a puff of smoke and rain back down to earth in a million pieces. This happened again and again.

One time Ted Taylor invited mathematician Richard Courant to come out and observe the flight tests. Courant, one of the greatest mathematicians of the day, was from Germany. He had worked with David Hilbert back at Göttingen, and he was an expert on shock waves. So Courant came out one Saturday morning and watched the test. The bomb went off and Hot Rod lifted up off the launch tower, and then—*kaboom!*—it disintegrated in front of his eyes like a firecracker.

"Zis is not nuts," Courant said in his German accent. "Zis is supernuts."

Finally, though, the Orion models flew—up to a few hundred feet anyway—but by this time Dyson had left California and gone back to the Institute. One day back in Princeton he got a letter from his friends who had stayed with the project. "Wish you could have been with us to enjoy the Point Loma festivities last Saturday," the letter said. "The Hot Rod flew and flew and FLEW! We don't know how high yet. Ted, who was up on the side of the mountain, guessed about a hundred meters by eyeball triangulation. Six charges went off with unprecedented roar and precision. . . . The chute popped exactly on the summit and it floated down unscathed right in front of the blockhouse."

But that was the last of the H-bomb spaceship project. The U. S. government decided to use chemical rather than nuclear rockets for its space program, and later, after the 1963 Test Ban Treaty had made nuclear explosions in the atmosphere and in space illegal, Project Orion was doomed forever. Hot Rod, though, still survives. It's on display at the National Air and Space Museum, in Washington, hanging by guy wires and pointing in the general direction of the Washington Monument.

And then there are the aluminum shards. Dyson used to walk around Point Loma after the tests, picking up fragments of the shattered spacecraft. He still keeps a few of them in his desk drawer at the Institute, in a clear plastic Baggie. They remind him of the time he almost made it to the stars.

When Lenard went in to see Dyson about the stability of matter problem, Dyson immediately got to thinking. "The point was," Dyson says, "that atoms can be so extremely complicated when there are a lot of them together. They can do all kinds of extremely weird and sophisticated things. Like they can be liquids or solids, or they can be all sorts of unstable chemicals and explosives and all kinds of things—and it's all done with plain atoms. Just ordinary atoms. And so the question was, how can you really be sure that with all this variety of behavior that matter is capable of, that a piece of it just couldn't collapse? That was the problem."

The solution wasn't apparent right away, to Lenard, Dyson, or anyone else. Lenard's office at the Institute was next to C.N. Yang's in building D. This was the same Frank Yang who had, with T.D. Lee, won the Nobel prize for physics for establishing the nonconservation of parity in weak interactions. Every now and then Yang and Lenard would bump into each other in the hallways and say hello, but Lenard—the junior man—never took it upon himself to walk into Yang's office and talk things over.

Which is strange. This is supposed to be one of the main purposes of the Institute—it's where the young come to study and learn at the feet of the masters—but sometimes there's very little of it going on. Dyson himself, for example, never went in to see Einstein or Gödel. "I knew Gödel in day-to-day life reasonably well," Dyson says, "but I never sat down with him and had philosophical discussions. I was always much too shy. Same with Einstein. I never went in and said 'Hello, Einstein, I want to talk to you.' What the hell should I waste his time for, you know? They obviously had more important things to do than talking to me."

Lucky for Andrew Lenard, then, that one day Yang should happen to knock on his door to come in for a chat. Yang wanted to know what Lenard was working on, and so he tells him about the stability of matter problem. Yang thinks this is curious: "Very interesting," he says. "That's either a trivial problem or a very difficult problem."

So Yang goes back in to his office—which is right next door—and Lenard starts to hear this *tapping* on the wall. He realizes it's Frank Yang writing on his blackboard. This goes on for a while—*tap, tap, tap, tap*—chalk banging against slate, and Lenard thinks no more about it.

Then all of a sudden the tapping stops, as if the poor man had a heart attack. Dead silence.

A few minutes later Yang pops his head back into Lenard's office: "It's difficult," he says, then vanishes.

The problem was not easy for Dyson, either. "He took, I would say, a couple of weeks," Andrew Lenard recalls. "Then he came back in to me and said, 'Look, it's a very interesting problem, and we can do this and we can do that, and we can estimate this and estimate that.' He had all sorts of different ideas, and I began to work a few of them out."

It so happened that Dyson and Lenard were on opposite ends of the campus, but there was a seminar room above Dyson's office, and the two of them would meet there for a few hours every day to thrash things out.

"There was a big blackboard in the seminar room," Lenard says, "and usually nobody else was around, and so Dyson used the blackboard and he would lecture to me for a while and then I would sort of, you know, challenge him on some of the points. Some of his arguments were not quite correct, not quite rigorous, but he just kept coming back with new ideas, lots of ideas. We would throw them around for a few hours, maybe, and then I would go away and think about them. I actually made very little progress myself, because by the time I figured out what Dyson was getting at, he would be coming up with some more ideas.

"Anyway," Lenard continues, "there were lots of technical difficulties, and this process went on for a while—two months, perhaps—and finally by this time everything fell into place."

Convinced that Dyson had pieced his way through the problem, Lenard advised him to publish the result, his explanation for why matter wouldn't go up in smoke. Dyson, however, insisted on joint authorship: it was Lenard's problem to begin with, and Lenard had done a lot of the work. "He was very generous about it," Lenard says. "But Dyson was the one who came up with the ideas, absolutely. No question about it."

And so, like many a paper written by a junior and a senior scientist, the junior man went away to write up the ideas of the senior partner. The paper, which turned out to be forty pages long in manuscript, was published in two parts, in the *Journal of Mathematical Physics*, in 1967–68.

So why is matter stable? "The answer," says Dyson, "is not awfully easy to summarize. The proof involves some clever mathematical tricks. But basically, you need the exclusion principle. The reason that matter holds itself apart is that the electrons always satisfy the exclusion principle."

The exclusion principle, stated by Wolfgang Pauli, holds that no two fermions—a subclass of elementary particles—may occupy the same quantum state. It's this mutual exclusiveness that holds the atoms apart and keeps them from falling into each other's empty spaces. "This result," Dyson says, "perhaps gives some philosophical understanding of why God had to invent the exclusion principle before he could create a world with matter in it."

As clever and full of technical tricks as it was, the Dyson-Lenard proof was shortly superseded. "It was not what would now be considered an elegant or a reasonable solution," Lenard says. "Even though there were many clever ideas in it, they were really thrown together in a way that was not terribly natural for an understanding of why it all worked."

"The technical part of it is completely obsolete now," says Dyson. "It was done much better later on by Eliot Lieb and Walter Thirring. The idea is the same, but Lieb and Thirring improved the mathematics so much that they did in four pages what we did in forty."

The Dyson-Lenard paper was not the standard type of thing found in physics journals: it was an explanation of something that physicists knew to be true all along, that matter was inherently stable. Most mainstream particle physicists, in the meantime, had their hands full just trying to cope with the new and unheard-of elementary particles that were streaming forth from cyclotrons like fireballs from Roman candles.

As late as the 1930s, the great ancients—Einstein, Bohr, and the rest—were contemplating a world composed of just a few basic entities: electrons, protons, neutrons, and photons, and they were looking forward to the day when they would be able to fashion a complete and consistent theory of the entire physical universe in terms of just these particles. The more deeply experimental physicists looked into the fine structure of matter, though, the more particles they found. For a while, the number of particles remained small enough so that you could still remember them all in your head, but then there came a time—back in the early 1960s, it was—when there were so many particles that physics journals took to printing wallet cards listing them all for you. Nowadays even a wallet card isn't enough: it takes the *Reviews of Modern Physics* a whole issue to print the full list of known particles, resonances, and other states of matter.

Theorists and experimentalists seemed to be in a constant race to keep up with the explosion of new particles. On rare occasions, theory would be ahead of experiment: theorists would predict that there would be a particle of such and so characteristics, and experimentalists would find it in their machines. Other times, though, particles would show up that no one had anticipated. I.I. Rabi's response to one such case—the μ-meson—is representative: "Who ordered *that?*" But keeping up with the particles was a race against nature, and like a succession of torch bearers, the Institute's theoreticians were carrying the fire.

Almost all of the great particle theorists came to the Institute for Advanced Study at one time or another: Albert Einstein and Niels Bohr, of course, but also Max von Laue, and I.I. Rabi. Later there were the Young Turks: Murray Gell-Mann, Hideki Yukawa, Sin-itiro Tomonaga, Aage Bohr, Abdus Salam, Frank Yang, T.D. Lee, Nobel laureates every one. Along with them were the rest of the makers of the quantum physics revolution: George Uhlenbeck, Robert Mills, Freeman Dyson, Abraham Pais, John Wheeler, Frank Wilczek, Geoffrey Chew, Bruno Zumino, Yuval Ne'eman, Gabriele Veneziano, Yoichiro Nambu, Tullio Regge, Marshall Rosenbluth, and many more besides. Two of the greatest, and surely the oddest, of all the Institute's particle men were P.A.M. Dirac and Wolfgang Pauli.

P.A.M. Dirac was the very picture of the reclusive genius, reticent to the point of monkishness. "My father made the rule that I should only talk to him in French," Dirac once recalled. "He thought it would be good for me to learn French in that way. Since I found that I couldn't express myself in French, it was better for me to stay silent than to talk in English. So I became very silent at that time."

And he seemed to stay silent forever afterward. Two Berkeley physicists once sat down with Dirac for a solid hour and presented some of their work in the hope that the celebrated man would venture a few critical comments. Their hope was in vain. At the end of the hour there was no response at all from Dirac. When the silence in the room began to get uncomfortable, he asked, "Where is the post office?," and left to buy some stamps. On another occasion Dirac was asked how he liked the novel *Crime and Punishment*, which he had read. "It is nice," Dirac replied, in a masterpiece of succinctness. "But in one of the chapters the author made a mistake. He describes the sun as rising twice on the same day."

Withdrawn from the world of things, Dirac would retreat to the world of equations. "A great deal of my work is just playing with equations," he once said. "I don't suppose that this applies so much to other physicists; I think it's a peculiarity of myself that I like to play about with equations, just looking for beautiful mathematical relations which maybe don't have any physical meaning at all. Sometimes they do."

One of the equations that did turn out to have a physical meaning gave a result that was more than he'd bargained for. Now called "Dirac's equation," it opened the door to a new world of particles, collectively termed *antimatter*. The equation described the behavior of an electron with almost perfect accuracy, but it also predicted an electron of a wholly new type, one whose charge was positive.

At first, Dirac thought that his equation was describing the proton in some sort of roundabout fashion, for the proton's electric charge is opposite to that of the electron, and, as he said later, "At that time . . . everyone felt pretty sure that the electrons and the protons were the only elementary particles in Nature." The positively charged particle his equation described was not a proton, however, for a proton's mass is almost 2,000 times greater than an electron's. If Dirac's equation meant anything, it meant that there existed in nature "a new type of particle, unknown to experimental physics, having the same mass and opposite charge of the electron." Dirac called it an "anti-electron."

A year and a half after he'd predicted its existence, the "anti-electron" turned up in a cloud chamber experiment run by Carl D. Anderson, of Caltech. Anderson called them "positive electrons"; today they're known as positrons. Dirac's equation applied to the proton as well as it did to the electron, and so it also described antiprotons, which—some twenty years after they were predicted—were duly discovered in the laboratory. In fact, the Dirac equation gave rise to a whole new realm of antimatter, and physicists had a lot of racing to do to catch up with the phenomena in front of them. "I think that this discovery of antimatter," Werner Heisenberg said later, "was perhaps the biggest of all the big jumps in physics in our century."

Dirac came to the Institute for an initial year in the 1930s, then returned once each decade right up through the 1970s, making him one of the place's more regular and long-term visitors. As reclusive in Princeton as he was elsewhere, he had a special fondness for the Institute's woods, into which he would disappear late in the afternoons with an ax in hand, muttering something about blazing a trail to Trenton. What with his silences and penchant for solitude, Dirac was temperamentally an Institute man through and through.

And, with an extreme preference for theory over experiment, for theories which "have to have a great mathematical beauty," he was also in the true Platonic-heavenly mold. "One can feel so strongly about these things," he said toward the end of his life, "that when an experimental result turns up which is not in agreement with one's beliefs, one may perhaps make the prediction that the experimental result is wrong and that the experimenters will correct it after a while. Of course one must not be too obstinate over these matters, but still one must sometimes be bold."

In terms of personality, never was there an Institute man more bold than Wolfgang Pauli. Physically a rather large specimen, Pauli bubbled over with a nervous energy that expressed itself in the form of some unorthodox bodily movements and offensive speech patterns. He had this habit of rocking back and forth on his feet, at the same time shaking his head from side to side, giving the impression that his muscles were somehow overstrung. Verbally, he was particularly good at the swift put-down. He once described an upstart physicist as "so young and already so unknown." If he didn't like someone's idea or theory he'd say it is "not even wrong." Even when he had a good word to say, which was infrequent enough, he would be sure to couch it so that it came as a slap in the face, as when, in a seminar given by Einstein, Pauli—who was just a graduate student at the time—stood up and allowed as how, "You know, what Professor Einstein says is not so stupid."

He also had a swelled head second to none in the physics business, and perhaps even in the whole history of science. Pauli once complained to Abraham Pais that he was having a hard time finding a new physics problem to work on. "Perhaps," he said, rocking back and forth, "that is because I know too much."

He used to put people down at physics conferences whenever they weren't being clear or correct enough for his taste. This happened to Oppenheimer once, at a seminar in Ann Arbor. Oppie was lecturing and covering the blackboard with equations when Pauli jumped up, grabbed an eraser, and cleaned the whole blackboard off, saying it was all nonsense. Pauli did this twice more, and probably would have continued, but finally Hendrik Kramers stepped in and told Pauli to sit down and shut up, which he did.

That was back in the 1930s. Twenty years later, and now at the Institute, Pauli hadn't changed. This time Oppenheimer was in the audience and the lecturer was Frank Yang, who was giving a talk on gauge invariance. Yang had barely started when Pauli interrupted him with, "What is the mass of this [particle]?" Yang replied that it was a complicated problem and that he hadn't come up with a definite answer. "That is not sufficient excuse," Pauli said. Yang, always the model of politeness and reserve, was so stunned that he had to sit down and collect himself. The next day, Pauli left a note in Yang's mailbox: "I regret," Pauli said, "that you made it almost impossible for me to talk to you after the seminar."

People stood for this kind of silliness because Pauli was, of course, a brilliant fellow. He had come out with the "exclusion principle," one of the pillars of the new physics, when he was just twenty-four. And, like Dirac, Pauli also added something new to the roster of elementary particles. The "Pauli particle," as it was called for a while, emerged through his analysis of beta decay, a form of radioactivity in which the nucleus loses an electron. At the time, this phenomenon was quite mysterious. For one thing, the nucleus doesn't contain any electrons to begin with, so how could it give any up? For another thing, physicists couldn't account for all the energy that was lost during the beta decay process. The nuclear-decay products—then called "beta rays"—were observed to carry off less energy than the atom itself released in the process. What happened to the rest of it?

The missing energy problem was so baffling that Niels Bohr, for one, was ready to abandon the hallowed principle of conservation of energy: in beta decays, he claimed, energy might not be conserved. Pauli, though, had a different idea. He speculated that the extra energy went off in the form of a new particle, one that experimentalists had never seen. The particle would have no mass and no charge; it almost wasn't there at all. Now this is an *ad hoc* way of dealing with phenomena—when you run into a difficulty, make up a new particle—and even the great Pauli was for a time too embarrassed by his suggestion to publish it. Later on, though, he got up the courage.

In the end, Pauli turned out to be right: his "invisible" particle is the neutrino. (Enrico Fermi contributed the name, which means "little neutron.") What actually happens in beta decay is that a neutron simply disintegrates into other things, into a proton, an electron, and a tiny, uncharged particle, the neutrino. Because it was uncharged and so small, the neutrino was not seen experimentally until 1956, although Pauli had the idea back in 1930. In the intervening quarter-century there was a lot of talk and worry about—as Oppenheimer once put it—"the haunting ghosts of neutrinos."

Pauli spent the war years—1940 to 1946—at the Institute for Advanced Study, working on the problem of mesotrons, or "mesons" as they're known today. (Like antimatter and the neutrino, mesons were another case of physicists' anticipating nature. Originally proposed by Yukawa in 1935, they were not observed in the laboratory for more than a decade afterward.) Pauli, first and foremost a theoretician, wondered if he shouldn't be going off to war like everyone else at the Institute: other than for himself, Einstein was the only physicist not directly involved in some kind of war work. He confided his worries to Oppenheimer, then at Los Alamos. Oppie told him to stay put, so that "when the war is over there will be at least some people in the country who know what a mesotron is."

In December of 1945, Pauli was awarded the Nobel prize for the exclusion principle. At the Institute dinner party celebrating the award, Einstein—who had nominated him for the prize—delivered a speech in praise of his colleague. Pauli, for once, could afford to be humble.

Today there are no more Paulis, no more Diracs, Einsteins, Oppenheimers, or Bohrs at the Institute for Advanced Study. The ancient greats have died, the Nobel prize winners have left, and—in the view of outsiders, anyway—the Institute's once-great particle physics program has become a shadow of its former self. "In particle physics," says Harvard's own Nobelist Sheldon Glashow, "the Institute certainly has declined. Their last permanent appointments in particle physics were Adler and Dashen, twenty years ago. An institution that doesn't make permanent appointments more often than twenty years is obviously moribund."

Would Glashow himself care to be at the Institute? "Oh, I wouldn't imagine going there," he says. "So far as I can see right now, the Cambridge area is the number one place for physicists."

Glashow isn't the only one not knocking down the Institute's doors. There's a long list of people who have refused offers of appointments at the Institute: "It's unbelievable what a long list that is," says Freeman Dyson. Great physicists seem to turn up their noses at the place, partly because of the lack of other great physicists there. "There's nobody there who'd attract me to go there for a year, let's put it that way," said Cornell's Michael Fisher.

But if the Institute is no longer brimming with particle physicists at the top, its program for training the younger generation still survives intact. "The main virtue of the Institute," says Sheldon Glashow, "is that it brings very talented young people to the Princeton area, and they, together with the established—and healthier—group at the university, have been doing good physics."

Like their brethren in astrophysics, the Institute's elementary particle physicists get together with their counterparts at the university for a weekly lunchtime seminar. The announcements go up on bulletin boards:

Princeton University—The Institute for Advanced Study

MONDAY LUNCHTIME SEMINAR
Monday, April 21, 1986

"Instantons and the Super Yang-Mills Beta Function"
Tim Morris
The Institute for Advanced Study

IAS Board Room, Lunch at 12:30 p.m. and the talk at 1:00 p.m.

Today's the day, and now's the hour—12:30 P.M.—and there are about a dozen particle physicists at a table in the Board Room, having lunch. Tim Morris is there, of course, and Mark Mueller, and Corinne Manogue—the lone woman particle physicist this semester—together with a bunch of others. They're not discussing physics; in fact, their talk is about politics, about the U.S. bombing strike against Libya, which took place just a few days earlier.

"Bombs are cheaper than chickens," one of them says. It's unclear what this means, but everybody laughs anyway. Nobody's defending Ronald Reagan.

Singly and in small groups others lope into the room carrying their lunch trays, and by 12:50 the rest of the group seems to have arrived. Two of the three Institute particle physics faculty are here: Freeman Dyson and Stephen Adler; Roger Dashen is elsewhere. There are some graduate students from the university, and a few faculty members. By now the talk has strayed back into science, and you can hear people saying things about compactification in SU(3) and—as always—about computers, about getting time on a Cyber 205, on the Illinois Cray, and on a Cray XMP somewhere else.

Julian Bigelow walks in at 1:00 o'clock on the nose—the same Bigelow who was chief engineer on the Institute's Electronic Computer Project, back in the good old Johnny von Neumann days. A little rumpled-looking in his red basketball sneakers and blue parka, Bigelow takes a seat front and center and spreads out a yellow legal pad in front of him, as if he's going to take everything down word for word. Good luck.

Corinne Manogue gets up from her chair, steps to the front of the room, and leans on the end of a dining table. She murmurs a few words—the sound level in the place has now risen to full din, with people talking and laughing and clinking their teaspoons against saucers—and nobody seems to pay any attention to her, but she introduces Tim Morris anyway, strictly business, just his name and the topic: "Our speaker for today," Corinne Manogue says, "is Tim Morris, who will talk on instantons and the super Yang-Mills beta function."

Somehow, people have heard this—or maybe it's the sight of someone standing at the front of the room that's the signal—but anyway they shut up and begin rearranging their chairs so that they can get a good look at the blackboard.

It soon becomes clear just why. At the astrophysics luncheons, speakers hardly ever use the blackboard, for the very good reason that they generally talk about *things in the sky* rather than about equations. In particle physics, by contrast, there's really nothing to talk about apart from equations. It's as if the thing-in-itself—the particle in question—*is* no more and no less a formula, and so these seminars are about one thing, and one thing only: *equations*. Tim Morris, the speaker, has come prepared. He's got what looks like about twenty sheets of paper in his hand, and all they seem to have on them is . . . *equations*.

Morris is a short, slight, British fellow with black hair and very prominent eyebrows. He's wearing a yellow sweater and tan corduroy pants. A journey of a thousand miles begins with a single step, and so Morris begins with a single equation, a short one, only a dozen symbols or so. This is mere warm-up, batting practice, as it were, getting into the waters. He writes the equation out on the blackboard, and he goes through it symbol by symbol, identifying and explaining every component: ". . . and this of course is the conformal fermionic zero mode," and so on and so forth. In the audience—which is absolutely silent, all ears—a few people are taking notes, but most of them just sit idly by and watch.

Morris writes more equations, explaining things as he goes along: "The infinite corrections arise in such a way as to renormalize the power of g here," and everyone takes this in without batting an eyelash, as if it were all just so much baby physics.

He writes more equations, meanwhile talking about "SUSYs" (supersymmetric theories), instantons, superfields . . . when suddenly there's an interruption, a question. It's a Princeton University faculty member back there in the audience, and there's a certain amount of disdain in his voice, a certain incredulous "this-is-too-much" inflection he puts into it: *How can you expect to get interesting solutions*, he asks, *when you're not taking account of the Legendre transforms*?

But Tim Morris has this well in hand. He's absolutely unruffled. "Unless there's something I'm just not seeing," Morris says, "I don't find any difficulty here." He calmly explains why he can get interesting solutions regardless of the Legendre transforms. The questioner doesn't look too happy about this, but he lets it pass anyway.

The seminar lasts for a solid hour. Equations go up on the blackboard, come down on the eraser, and more go up. Mostly, Morris speaks from notes, but now and again he pauses and steps back and summarizes what he's done so far. Every so often he draws a picture—or something that could be a picture if it pictured anything. It looks something like a poached egg, more like a formless blob.

A little after 2 o'clock Tim Morris puts up the last of his thirty-five equations. Abruptly, he stops talking, turns to the audience, and raises his big eyebrows an inch or two. This means The End, and so people applaud, quite loudly. Not one person in the audience has left the room the whole time, even to get dessert. Their eyes have never left the *equations*. Now, with no more *equations* to take in, and apparently no more questions—everyone leaves in a rush.

I meet Freeman Dyson in the stairway of building D the next day. "How did you enjoy our little seminar?" he asks.

"Well, it was Greek to me," I tell him.

"Then you understood as much of it as I did," Dyson says.

"Yeah, well . . . I even went over and talked to Tim Morris afterward."

"Oh," Dyson says, "then you understand it much better than I do."

Tim Morris is twenty-six years old. He went to Cambridge University as an undergraduate, then to Southampton University for his graduate work. He was there for three years, then came directly to the Institute for Advanced Study. He's in this country on a Harkness fellowship, the same kind of fellowship that Freeman Dyson had when he came to the United States in 1947. "It's some millionaire who wants to get people over here to see how wonderful America is," Tim Morris says.

"So how is it?"

"It's great," he says. "Love it!"

Tim Morris worked on super Yang-Mills theories for his Ph.D. degree, and now he's more or less a world-class expert on the subject. While writing his doctoral dissertation, he had gotten interested in the work that some Russian physicists had done on instantons. An instanton, Morris explains, refers to a tunnelling process that takes place between two force field configurations. "It's called an instanton because in practice this thing—the tunneling—would take place in an instant of time," he says. "Just like that."

Originally, Morris and his thesis advisor, Douglas Ross, thought they could show that the Russians' argument was correct, but they ended up demonstrating just the opposite. "The whole point of my seminar was really that we went out to try and prove them right, and in the end we believed we proved them wrong. We ended up doing all that work just to put nails in their coffin."

I ask him if this happens very often: getting negative results, working on things that in the end don't turn out to be right.

"Theoretical physicists," Tim Morris says, "expend a great deal of energy in disproving what another one says, or trying to disprove the other one, or trying to build on another's theory, whether or not that theory has anything to do with nature at all. Like the theory I talked about today: it's nothing like nature because it's too simplistic a theory. And most of the theoretical physics that goes on is about theories which the physicist in fact *knows* don't in fact correspond to nature. A lot of energy has been expended on two-dimensional theories, for example. Lots of people did this. But we don't live in two dimensions, we live in four. So what's the point?"

Good question. "What *is* the point?"

"If you want to be cynical," Morris says, "the reason people have worked in two dimensions is that it's *easier* to do two-dimensions than to do four. And if a load of guys on one side of the world start working on two-dimensional theories, then a group on the other side will look at it and say, 'Well, they missed this point, they could have done this,' or 'They did that wrong,' or whatever, and once this gets going then you can have hundreds of papers being published on stuff that has nothing to do with the way nature really is."

"Shouldn't people be working on theories that are right, instead of trying to disprove all the wrong ones?"

"But we don't *have* the true theory of nature," Morris says. "And as long as you don't know what the true theory of nature is, then it's a good idea to work on all possible theories because one of them may in fact turn out to be the right one after all. Perhaps two-dimensional field theories is even a good example. I was really scathing at people who worked in two dimensions because as far as I could see the only reason they did it was because it was easy, but then along comes string theory—and strings are two-dimensional surfaces—and suddenly all that two-dimensional field theory has an application. It makes sense."

Particle physics works this way, Tim Morris says, because the field is so complicated.

"It's very hard to write down a theory of something and be right, completely right. Theory is all about making things simpler than they really are. In Newton's day, you dealt with one, two, or three particles. In reality you've got billions and billions of particles all interacting together. Nobody could calculate something like that."

"But if things are so complex, then how can you have any confidence in the ultimate rightness of your theory?"

"Because you've got loads of guys out there who are all too willing to shoot you down if you're wrong."

There's never any shortage of people willing to shoot you down if you're an Institute particle physicist. In fact, some of them find the whole experience to be a mixed blessing. "If I'm looking for a job somewhere, the Institute will count for a lot," says one of the younger members. "And the proximity to the university is good in many ways, because you're next to some really hot guys. On the other hand it's a shame because it really builds the pressure up and it makes the place not so nice to live in. It must be much worse at Princeton than it is here, though—the Institute itself is quite laid back."

"I don't understand. What's the source of the pressure?"

"The source of the pressure is to do as well as the guys at Princeton. If you don't do as well, then the Princeton guys are going to tear you to shreds."

"Don't you want the best criticism possible, though?"

"Yeah. But there's a difference between criticism and being ripped totally apart at seminars. That's happened. As a power trip on their part, to show you how great they really are. Tim Morris was lucky at the seminar. He didn't get ripped apart at all, although I thought he was going to be. The good guys at Princeton do have the reputation for ripping lesser physicists apart."

I ask him if the Institute is becoming overshadowed by its next door neighbor.

"At this moment in time," he says, "I don't think the Institute is doing anywhere near as well as in its past history. And this shouldn't be. It shouldn't be that Princeton University is the great shining star and that we're just a complete shadow to it, which is what's happening at the moment. It's a pity that the permanent members here . . . Well, everybody does burn out at some stage, so it's no criticism of them, really, but the permanent members here do seem to have burnt out. And the Institute itself, I don't know what it's doing. Their policy seems to have backfired a little. They've been waiting for too good a guy, I guess, and as a result they haven't got anybody. There are no leading lights here, which is a shame."

The Institute's last leading light went out when Oppenheimer died. Oppie was larger than life, he was glamorous, and people are attracted to glamour, even in the rotting-tweed world of academia. There's no one glamorous left at the Institute, although there easily could have been. For a while after Oppie stepped down there was some talk that Murray Gell-Mann might be made director. Gell-Mann is glamorous: a Nobel prize winner, alarmingly smart, and no shrinking violet. Oppie's own choice for his successor—Frank Yang—declined the position, and when in 1965 Oppenheimer announced that he would retire the following year, the Institute decided it was time to take stock. It had now been in existence for thirty-five years, a physicist had headed it up for the last eighteen, and perhaps another one ought not to be appointed.

The Institute's board of trustees therefore created from its ranks a Committee on the Future of the Institute. After six months of work, the committee decided the time was ripe for new directions . . . which was exactly what the faculty had always lived in mortal fear of. They hated change, and with good reason: when things are perfect, as they must be in a Platonic-heavenly Paradise for Scholars, then any change is necessarily for the worse. The full board of trustees nevertheless accepted the committee's recommendation that the Institute ought to immerse itself in the problems of contemporary society, and they hired as director Carl Kaysen, who they were sure was equal to the challenge.

Born in Philadelphia in 1920, Kaysen had gotten his Ph.D. at Harvard and then went into public service. He worked for McGeorge Bundy in the Kennedy White House, went to India with Averell Harriman, and then to Moscow, to work on the Limited Nuclear Test Ban Treaty. To Institute regulars, however, none of this altered the fact that Carl Kaysen was not at all in the Institute mold. For one thing, he was an economist, and economics is not what could be called a "science," except by a very charitable extension of the word. In any case it was a discipline too rooted in the dirty empirical realities to be entirely welcome at the Institute. (Oppenheimer himself had, quite early on in his directorship, merged Flexner's ill-fated School of Economics and Politics with the School of Humanistic Studies.) Then again there was the matter of Kaysen's books: *United States v. United Shoe Machinery Corporation: An Economic Analysis of an Anti-Trust Case* (1956); *The American Business Creed* (1956); *Anti-Trust Policy* (1959); *The Demand for Electricity in the United States* (1962). Institute regulars took a look at these titles and their eyes began to glaze over. As André Weil put it, "I think he wrote his thesis about a shoe factory."

Notwithstanding all this, Kaysen was made the fourth director of the Institute in 1966. At the beginning, he got on well enough with his faculty. The physicists had been dribbling away so Kaysen hired four new ones in replacement, particle physicists Stephen Adler and Roger Dashen, plasma physicist Marshall Rosenbluth, and astrophysicist John Bahcall. Kaysen also put into effect his plans for a School of Social Science and experienced no opposition from the faculty when he appointed anthropologist Clifford Geertz, a Harvard Ph.D. with five books to his credit.

Two years later, though, the roof fell in on Carl Kaysen. He nominated Robert N. Bellah for a permanent professorship. Bellah was a sociologist, and to the Institute regulars—especially the mathematicians—sociology was like fingernails on a blackboard. Mathematicians live in a world of eternal truth and crisp perfection . . . which are not exactly the qualities that anyone associates with sociology. "Many of us started reading the worthless works of Mr. Bellah," mathematician André Weil confessed. "I've seen poor candidates before, but I've never had the feeling of so utterly wasting my time." But contempt for Bellah was not confined to the mathematicians. Harold Cherniss, a classicist and philosopher, observed that, "It was clear as crystal that Bellah was not of the intellectual and academic quality of a professor at the Institute."

Bellah had written one full-length book, on Tokugawa religion, dozens of articles, and had collected some of his more personal, less-scholarly essays in a volume called *Beyond Belief*. To the Institute regulars the book's title all too accurately described their impressions of its contents. To them, Bellah's highly emotional approach to things was a blast of chill air, and at the faculty meeting called to consider the nomination, Institute stalwarts voted Bellah down thirteen to eight, with three abstentions. Kaysen, however, said he was going to appoint Bellah anyway. This was the last straw for the Institute faculty. Not only wouldn't they hire Bellah, they'd get rid of Carl Kaysen too.

This led to Faculty Mutiny Number Three.

Of all the faculty revolts at the Institute for Advanced Study, this was by far the biggest. It was a major scandal, and made for some sensational reading in the pages of the *New York Times*.

"We do not trust Kaysen's judgment, his fairness, or his word," Deane Montgomery was quoted as saying. "He is essentially a politician with almost no interest in or appreciation for advanced study. He is eager for power but does not have the moral integrity or intellectual capacity to use it wisely."

(Others, meanwhile, were wondering what had gotten into Deane Montgomery, who had opposed Oppenheimer, and was now opposing Kaysen. "I always thought of Deane Montgomery as a wonderful man," said one Institute member. "But at the Institute he behaved like an ogre for some reason, trying to devour directors.")

"Kaysen's usefulness as a director is at an end, and the sooner he realizes it, the better," mathematician Armand Borel was quoted as saying.

"There are seventeen people who have lost confidence in the director. That the Institute can function under these circumstances is unthinkable. Our lack of confidence is such that we don't see the point of any agreement with Dr. Kaysen, since he might very well put it aside at his pleasure."

(Why were the *mathematicians* the ones who were fomenting all this trouble? One member explains: "You know what they say about mathematicians, that everything they can get done in a day—because it's so concentrated and takes so much out of them—they can do in a few hours in the morning and then they've got the rest of the day to bug other people.")

Kaysen, though, decided he was going to gut it out, at least for a while. "I'm intending to stay and serve as long as I can usefully do so," he said in 1973. He did, after all, have supporters on the faculty, and not only in the person of Clifford Geertz, whom he had hired. Freeman Dyson, for one, thought then, as he does now, that Carl Kaysen was the best director the Institute ever had. "He didn't stand in awe of anyone," Dyson says. "He went around asking tough questions, like *Is this place really all it's cracked up to be*? He gave us all a kick in the pants."

But two years later Carl Kaysen quit the directorship and left the Institute. He said, in a letter addressed to his faculty, "Ten years of academic administration and entrepreneurship are enough; and I wish to spend the next decade or two in more agreeable ways."

It so happened that mathematician André Weil, one of Kaysen's staunchest foes, was due to retire on the very same day that was to be Kaysen's last day as director. "When it was getting close to my retirement," Weil says, "I thought of asking the board of trustees for a special extension of my job, just for twenty-four hours. This was so I could enjoy one Kaysenless day at the Institute." In the end, Weil didn't go through with it. Robert Bellah, meanwhile, who had spent a year at the Institute as a temporary member, also left to return to his post at the University of California, making moot the whole matter of his appointment.

And thus ended Faculty Mutiny Number Three.

Whatever Carl Kaysen's misjudgments in the Bellah affair, everyone who's been at the Institute ever afterward ought to give Kaysen a silent nod of thanks for bringing the place fully into the twentieth century, not so much intellectually as physically. It was Abraham Flexner's notion that buildings and grounds ought to take a second place to people. "Brains, not bricks and mortar," was his motto, and his adherence to it was reflected in the aggressively utilitarian structures that he put up on Olden Farm. Like the good Platonist he was, Flexner seemed not to have much feeling for the way in which the sheer physical comeliness of their everyday surroundings could enhance the spiritual life of the Institute's "workers." Carl Kaysen, however, was not so grimly Platonistic.

"Ultimate standards in the intellectual world are aesthetic," Kaysen once wrote. "Terms such as originality, depth and elegance are now used in an approving way to characterize intellectual work. Thus it is appropriate to the Institute's purpose that it seek beauty as well as utility in the structures that house its activities, and embody it in visual form."

So Kaysen, who raised more than 8 million dollars for the Institute during his tenure as director, put together a building fund and constructed a new dining hall and a new office building for members. The old lunchroom used to be so crowded that members had to eat and run, simply for the sake of making another seat available, a practice that did not encourage the free exchange of ideas over a meal. The new structures, by contrast, brought a new sense of peace, spaciousness, and harmony to the Institute, and Kaysen was not just patting himself on the back when he wrote, in a report to his board of trustees, that "All have benefited from the sense of order and form which [the new buildings] provide. Not all have been explicitly aware of the source of their gratification, but nearly all have experienced it."

None of this was noticed by Kaysen's enemies, however, who lambasted his award-winning buildings as the most expensive per square foot ever built in the country. But to them Carl Kaysen could do nothing right.

Ten years after Kaysen's departure the buildings he put up are the jewels of the campus, and the School of Social Science is small but prosperous. The Institute's particle physics program, by contrast, bottom-heavy as it is with young members on temporary appointments, is looking to recapture some of its lost glory. There's still plenty of room at the top for a Dirac, a Pauli, or an Oppenheimer.

Chapter 9

The Truth About Things

A good scientist is not humble. Oh, you can be a shy enough fellow in the flesh, you can be meek and retiring and a social wallflower—just think of Kurt Gödel—but you can't afford to be intellectually bashful or timorous about your science. You've got to be bold, arrogant, perhaps even a little reckless. For you have to believe that by the simple application of your own mind to the facts of experience, you can discover the truth—a little part of it, anyway—the truth about nature and how it works.

At the Institute, scientists have beheld—in their mind's eye, in their equations, in their data—both the largest and the smallest structures of the universe, clusters of galaxies at the one extreme to unimaginably tiny distances of 10^{-291} centimeters on the other. One Institute member even went so far as to program a computer simulation of the entire physical universe. This is hubris.

Indeed, when you stop to think about it, it's a wonder that any of this stuff is even possible. How is it possible to know the extremes of creation? How is it that scientists, who are after all just ordinary human beings, how can they behold the incomprehensibly large and the preposterously small and come away with anything like knowledge? To imagine that a select band of human beings can, through the exercise of abstract thought, somehow escape their finite sizes, get out of their everyday six-foot-long frames, and be on speaking terms with the uttermost antipodes of large and small, this seems a minor miracle. Even some scientists think so.

John Bahcall, the Institute's senior astrophysicist, once gave a talk at Vassar College about the solar neutrino problem. During the question period Bahcall was asked about the current status of the oscillating universe theory, according to which the cosmos will ultimately collapse into a primeval fireball like the one from which it arose, that it will then explode back out again in another "big bang," and that this cycle of contractions and expansions will continue on forever. The question was whether astronomers still accepted this idea.

Bahcall mentioned the bumper sticker he had on his car—"The Big Bang Is an Exploding Myth," it said—but then shocked his audience with what is, for an astrophysicist, a rather puzzling admission: "I personally feel it is presumptuous to believe that man can determine the whole temporal structure of the universe, its evolution, development and ultimate fate from the first nanosecond of creation to the last 10^{10} years on the basis of three or four facts which are not very accurately known and are disputed among the experts. That I find, I would say, almost immodest."

How's that? An astrophysicist badmouthing presumptuousness? A master of the discipline in which light years and megaparsecs are everyday bread and butter, in which students of the subject regularly use little plus signs—or worse, *dots!*—to represent entire galaxies, he's saying that knowledge claims about these things are somehow . . . *bad form?*

But Bahcall went on to explain. Think of the ancient Greek cosmologies, he said. Think of how they used to believe that the universe was a ball on the back of a turtle. Probably a hundred years from now future scientists will look back at twentieth-century cosmological models and shake their heads in equal amusement and disbelief. "So I don't take those models very seriously," Bahcall said. "That doesn't mean that I don't publish papers based upon them, but it does mean that I tend to regard them pretty much the same way that I would regard Newton's theology. An interesting intellectual exercise, but nothing to be worried about."

Bahcall was putting his finger upon one of the great unresolved riddles of contemporary science, the problem of scientific truth. To wit: What is it that all this scientific thinking and theorizing and calculating get us in the end? What's the payoff of all our observation, quantification, computer modeling, and everything else? Does science (as we've always been taught as children) actually lay bare the facts of nature? Does it tell us *The Truth* about things? Or does it give us something far less exalted and pretentious, something on the order of "provisional hypotheses," "adequate interpretations," "heuristic guidelines," and the like? The matter is of no small consequence. The Institute for Advanced Study operates on a budget of well over 10 million dollars per year, and in the United States as a whole scientific research is supported to the tune of many billions of dollars, both public and private. What does all of it buy? Do we at least get part of *The Truth* in return, or are we spending these fortunes to learn things that are—at least as far as the ultimate truth goes—little better than Newton's theology or turtle-top views of creation?

"Many scientists never really worry about these questions," John Bahcall says. "They're not questions which affect their research or their temperament."

But if you nevertheless *ask* Institute scientists those questions, if you ask them point blank whether their theorizing gets them to the truth or only a convenient myth, the strange thing is that many of them can't decide what to say. Some, of course, insist that the entire goal of their work is to learn what the truth is, pure and simple. Thus, Institute physicist Stephen Wolfram: "I want to know the truth, however perverted that may sound."

Often as not, however, you'll get a two-stage response, on the order of: "Do I think that my explanation of why that galaxy is oblate rather than prolate is the *true* explanation? Well, yes, I guess so. Sure. . . . But you have to realize that this is only provisional. As for the 'ultimate truth,' well, . . . I guess I really don't know what you mean by that."

Thus, Institute particle physicist Steve Adler: "I think we do get knowledge, but I don't think we get final answers. Well, what you get are theories that work within a certain domain of validity. It depends on whether you say knowledge is absolute knowledge—you know, the final answer—or that we've got some approximation to what the final picture is. At least we have rules that allow us to calculate what happens in rather complicated experimental situations."

Thus Ed Witten, one of the inventors of superstring theory: "Well, whether we get to the truth or not, in any case we learn things which last. The view that older theories get overthrown is misleading. We learn new things. We develop more powerful laws, laws that unify principles and give us more accurate descriptions of more and more phenomena. It doesn't mean that the old stuff was wrong. It just wasn't complete."

But does Witten think that superstring theory, for example, is *right,* that it's the final word about nature?"

"Yes. When I say it's right what I mean is . . . it may turn out to be the ultimate theory, the complete theory of nature. But when I said it was 'right,' I merely mean that it's definitely a step forward."

John Bahcall has an explanation for all this back-and-forth waffling: "You have to distinguish between the way we talk and the way we think when we reflect on it. Scientists have a shorthand for talking which leaves implicit a lot of assumptions. We may well talk as if we're interested in 'the truth,' but actually this is just a shorthand for something like 'useful description' or 'better approximation.' "

If scientists themselves don't agonize over whether their work puts them in touch with the truth about things, some Institute members have spent the better part of their professional lives doing just that. These are the humanists, specifically the philosophers, more specifically still, the philosophers of science. The hot issue in philosophy of science in recent years is whether science provides us with true and genuine knowledge, or if not, then what *does* it give us? Naturally enough, there are two major viewpoints. One side looks at the history of science and, like John Bahcall, sees one theory falling at the hands of another, the ball-on-the-turtle theory being replaced by the geocentric theory being replaced by the heliocentric theory and so on and so forth. The conclusion here is that science doesn't offer up truth but only a succession of different interpretations. Some interpretations may be better than others for specific purposes—recent astrophysics being quite handy for making a soft landing on the moon, for example—but no one theory can be said to be unequivocally true and correct in any neat or final sense.

Those on the other side of the dispute are aware of the same historical facts, only they come to a different conclusion about them. They see primitive theories being replaced by more sophisticated ones, and they claim that some of these theories *have* indeed gotten to a final truth about nature. The world is not rolling around on a turtle shell, the earth is not at the center of the galaxy, but the sun *is* in fact at the center of the solar system. In other words, there's an ultimate truth out there waiting to be found, and it's the business of science to find it.

Today, though, the latter viewpoint is quite out of favor with philosophers of science, and, evidently, equally out of favor with at least some of the most distinguished practitioners of contemporary science. The prevailing philosophic attitude seems to be something like this:

Truth? Absolutes? Ultimate reality? Oh no, this is not the Dark Ages. It's not even the nineteenth century. This is modern times, and we know better. Better than to believe in anything like *The Truth*. We've been enlightened, emancipated, we've had our consciousness raised to the point where we can now face the fact that the only things that human beings can lay hold of are various alternative viewpoints, each of them correct in its own way. Even if there were such a thing as an "objective reality," there's no way to get out there to examine it directly. The reason is that we're all wearing rose-colored glasses, meaning that our previous experience, our culture, language, preconceived world views, and so on, all these things get in the way of our seeing the world as it is in itself. People used to think—back in the good old days—that we could behold nature *tabula rasa,* without preconception. This is an illusion. The sooner all of us own up to our rose-colored glasses, the better.

This, at any rate, is the viewpoint fostered by the guru of emancipated thinking about science, Thomas S. Kuhn. Kuhn first came to the Institute for Advanced Study in the fall of 1972, ten years after the publication of his book, *The Structure of Scientific Revolutions,* a blockbuster in the history of science. When the book came out, it caused a sensation. For one thing, it pooh-poohed the then-received view of scientific progress, according to which science brings us, in incremental stages, ever more closely to a final, impersonal truth. For Kuhn, science doesn't work that way at all. What really happens is that scientists look at the world and puzzle things out through their own collective set of rose-colored glasses, which is to say their common body of received opinion, shared assumptions, and expectations, both about their discipline and the world at large. Kuhn called these common beliefs "paradigms," and said that the important thing about them is that they tend to operate as blinkers and blinders, forcing scientists to see reality in certain specific ways as opposed to others.

For example: If a given scientific community accepts the idea that nature is alive, the notion that there's an *élan vital* at work in the universe, then those scientists will be inclined to interpret natural phenomena teleologically. They'll see events happening for purposes, and view them as if they're all part of a master plan. But another group, one which holds to a philosophy of mechanism (according to which events occur in strict "billiard-ball" cause-and-effect sequences), those scientists will perceive the world quite antithetically. As to whose picture of reality is *correct,* well . . . that's not a question we ask, Kuhn said. At least we don't ask such a question as scientists.

This view of science may have been controversial among scientists, but it was nothing but good news to the humanists. They loved it. Kuhn, after all, humanized science, he put *people* back into the scientific enterprise: science is not the cold and bloodless piling up of one impersonal abstraction on top of another, it's life itself, human theater, the drama of real people thinking real thoughts. That these thoughts are often preconditioned by prior experience—one's culture and language and so on—well, so much the better. It only goes to show that scientists are human like everyone else, and not a bunch of great shining gods who go out and discover *The Truth.*

So much for scientific immodesty and presumption.

Before long, *The Structure of Scientific Revolutions* was a book that everyone seemed to have read, and by 1969 Kuhn was one of the two or three most frequently cited authors in the United States. People suddenly began to see "paradigms" and "paradigm shifts" everywhere—almost as if it were a sport, an egghead recreation of some kind—and for a while your adroitness in juggling paradigmatic shifts and cosmic revolutions became an index of your meta-scientific enlightenment and degree of intellectual chic.

Kuhn, meanwhile, who had only wanted to see the progress of science correctly (without any rose-colored spectacles), was finding it hard to get any work done. He was on the Princeton faculty, teaching courses in the history and philosophy of science, and at the same time trying to get started on the history of quantum theory he wanted to write. But he was having the usual academic's trouble of finding the time. What with all the requests to give lectures, write articles, and reply to critics—which he had in no small numbers—Kuhn had little time left over for research and writing. He was, in other words, a perfect candidate for the Institute for Advanced Study.

In the spring of 1972 Kuhn worked out an arrangement with the university according to which he'd teach only one semester per academic year; he'd spend the rest of his time at the Institute.

Kuhn took up residence in the West building, the new office building put up by Carl Kaysen. Here he was in among the Institute's historians and social scientists, some of whom were made a trifle uneasy by having a physicist in their midst. They needn't have been so concerned, for although Kuhn had been trained as a physicist—he holds a Harvard Ph.D. in the subject—he had long ago decided to abandon theoretical physics in favor of the history of science.

"My own enlightenment began in 1947," Kuhn says, "when I was asked to interrupt my current physics project for a time in order to prepare a set of lectures on the origins of seventeenth-century mechanics."

Tracing his subject back to its roots, Kuhn read Aristotle's *Physics* and promptly came down with an acute case of paradigm shock. Aristotle was of course one of the greatest minds of antiquity, and you'd expect a lot of great science to be imbedded in his works, but the fact was that the man was just so . . . *wrong* about things, so flat out plain mistaken about the way the world works. It was puzzling.

How could Aristotle's characteristic talents have failed him so completely when applied to motion?, Kuhn asked himself. How could he have said about it so many apparently absurd things?

Among the absurd things Aristotle said was that all earthly objects were made up of four primary elements—earth, air, fire, and water—and that each of these seeks its own proper place in the overall scheme of nature. Earth, for example, seeks the lowest spot, water the next highest, air above that, and fire the highest of all; in this way the motion of all things is supposed to be explained, in terms of the natural tendency of each object to occupy its own special niche. Aristotle also believed that heavier objects fall more rapidly than lighter ones, that a ball thrown into the air keeps on going because the air behind it pushes it along, as well as a number of other doctrines that modern physicists regard as either quaint or perverse according to their custom.

To Kuhn, this comedy of errors was profoundly disturbing. Aristotle was such a genius—a sharp observer, an original thinker, the inventor of logic—that it was hard to believe that he was literally *mistaken* about all this stuff. But on the other hand there was no denying the fact that a lot of what Aristotle had said was just ridiculous. "The more I read, the more puzzled I became," Kuhn says. "Aristotle could, of course, have been wrong—I had no doubt that he was—but was it conceivable that his errors had been so blatant?"

Presently Kuhn was experiencing his enlightenment.

"One memorable (and very hot) summer day those perplexities suddenly vanished. I all at once perceived the connected rudiments of an alternate way of reading the texts with which I had been struggling."

The alternate way was to adopt Aristotle's *own* way of looking at things, to put on *his* rose-colored glasses. When you opened your eyes, lo and behold: there would be Aristotle's universe spread out before you in all its eldritch glory. All would be clear forthwith.

For Kuhn, the view through Aristotle's lenses changed everything. "I did not become an Aristotelian physicist as a result," he said, "but I had to some extent learned to think like one." Aristotle had been working within a paradigm that we no longer share, a schema in which qualities—such as place, shape, and purpose—were the primary realities. Once you accepted the notion of 'natural place,' what could be more logical than that every material body 'seeks' the location where it by nature belongs? Aristotle's once-strange mechanisms now seemed—in their own way—quite sound, his errors no longer so erroneous. "I still recognized difficulties in his physics," Kuhn said, "but they were not blatant and few of them could properly be characterized as mere mistakes."

From here Kuhn went on to postulate that all scientists work within paradigms, distinctive sets of commitments that they share with the other scientists of their era. A paradigm specifies what is and what is not acceptable as a theory, and in this way it determines the form and content of science. They're despotic, these paradigms, for they control what it is that scientists literally *see* when they look upon the world. It's like the visual gestalts at work when you look at one of those figure-and-ground pictures. First you see an unintelligible pattern . . . which then looks like a vase . . . which then looks like two faces looking at each other. Or you see something that looks like an up staircase . . . but if you look at it differently it seems to be a down staircase, and so on. Paradigms, of course, work on the grand scale, as if the world itself were a figure-and-ground illusion, ambiguous and amorphous in itself, and visualizable in as many distinctive ways as there are ways of looking at it. For Kuhn, what you see is a function not so much of what's *there,* but of the intellectual apparatus and excess baggage that you bring to the act of perception.

"Looking at a bubble-chamber photograph," he says, "the student sees confused and broken lines, the physicist a record of familiar subnuclear events. Only after a number of such transformations of vision does the student become an inhabitant of the scientist's world, seeing what the scientist sees and responding as the scientist does."

With this notion of science in hand, it was an easy matter for Kuhn to explain scientific revolutions: they occur when one paradigm is replaced by another. Mechanism replaces teleology; quantum-mechanical randomness replaces mechanism. Nature itself does not change: it's only our way of *seeing* it that does. Because people are so attached to their paradigms, scientific revolutions tend to be accompanied by a certain amount of intellectual bloodshed, and for the same reasons that political revolutions are: in both cases the underlying issues are not rational but emotional, and are settled not by syllogisms and rational analysis but by 'irrational' factors like group affiliation and majority rule. "As in political revolutions," Kuhn said, "so in paradigm choice—there is no standard higher than the assent of the relevant community."

It follows that scientists don't choose their paradigms on the basis of reason, but by something else entirely. "The issue of paradigm choice can never be unequivocally settled by logic and experiment alone," Kuhn said. "In these matters neither proof nor error is at issue. The transfer of allegiance from paradigm to paradigm is a conversion experience that cannot be forced."

To some, the most revolutionary, not to say disturbing, part of Kuhn's view of science and its progress was that it seemed to leave things like knowledge, truth, and external reality completely out of the picture. Indeed, in *The Structure of Scientific Revolutions,* the issue of truth enters the discussion only at the end of the book, where it surfaces almost as an afterthought. "It is now time to notice," Kuhn wrote, "that until the last very few pages the term 'truth' had entered this essay only in a quotation from Francis Bacon. . . . Inevitably that lacuna will have disturbed many readers."

But Kuhn only went on to confirm the reader's worst suspicions. In science, he said, truth is an optional and gratuitous concept. "Does it really help," he asked, "to imagine that there is some one full, objective, true account of nature and that the proper measure of scientific achievement is the extent to which it brings us closer to that ultimate goal?" Kuhn didn't think so. Paradigms themselves are neither true nor false: they are either held by a given scientific community or not held by that community, and that's about as far as it goes. You can of course explain why one paradigm replaced another: you can trace out all sorts of complex sociological relationships between paradigms and the communities that hold them, but the one thing you cannot do is to compare a paradigm with reality. To do that you'd have to see reality "as it really is," which is to say apart from any paradigm. But the whole point is that you *can't* see reality as it really is: scientists, like the rest of us, are condemned to see it only through their rose-colored spectacles.

What follows from all this is that science does not and cannot get hold of "the truth" in any objective and impersonal sense. And it follows that there is no such thing as progress in science—if what you mean by progress is getting closer to the way things are in themselves. "We may," Kuhn says, "have to relinquish the notion, explicit or implicit, that changes of paradigm carry scientists and those who learn from them closer and closer to the truth."

Just after *The Structure of Scientific Revolutions* was published, Thomas Kuhn attended a meeting of the Philosophy of Science Association in Cleveland, where he met a philosopher and historian of science by the name of Dudley Shapere. Like Kuhn, Shapere had gotten a Ph.D. degree from Harvard. Shapere's degree was in philosophy, though, rather than physics, and one might therefore have expected him to be solidly behind Kuhn, perhaps congratulating him for having made science safe for humanists. But in fact Shapere was appalled. Kuhn was a relativist, Shapere thought; he denied the objectivity and rationality of science; he implied that science is little more than collective subjectivism, mass delusion, a series of fads dressed up to look presentable. None of this was acceptable to Shapere, who believed that there certainly was an external reality out there, that science can grasp it, that the apprehension of objective truth was, if not the entire *raison d'être* of science, at least an essential part of the whole enterprise. Shortly afterward, Shapere published a review of Kuhn's book, describing it as "a sustained attack on the prevailing image of scientific change as a linear process of ever-increasing knowledge."

In the late 1970s Dudley Shapere also came to the Institute for Advanced Study. Here he worked out a view of science that would show how, despite the fact that they work in communities defined by shared assumptions and background beliefs, scientists can and do uncover genuine truths about nature. He wanted, as he said, "to remove barriers to the possibilities of human thought."

Shapere denied that seeing through rose-colored glasses makes a true understanding of nature impossible. If we know that we're wearing rose-colored glasses, we can correct for that fact and see nature in her true light. Shapere's term for Kuhn's concept of paradigm is "background beliefs." (One particularly assiduous critic of Kuhn examined *The Structure of Scientific Revolutions* and found that the crucial word "paradigm" was used in twenty-two different senses. "Clarification is obviously called for," Kuhn said in response.) You can be aware of your background beliefs, Shapere argued, and if you're a scientist you'll correct for those which stand in the way of the truth.

Religious, political, and metaphysical beliefs may interfere with a scientist's thinking. Isaac Newton, for example, based his cosmology on principles of the Christian religion. All this means, however, is that people aren't born with an instinct to follow scientific method. This is something you have to *learn*—you have to "learn how to learn," as Shapere put it. The fact is, though, that we *have* learned how to learn about nature, and the result of it is modern science.

The other half of the picture is the "given" which appears in experience. Even though we may be wearing rose-colored spectacles, there's plenty that shines through them unaffected. The colors of objects may be skewed to the red, but the shapes, sizes, textures, and other qualities out there come through as they really are. If we may not see nature's colors truly, still we can see everything else there is to be seen out there and we can see all those things correctly. Rose-tinted glasses may color our view of reality but they surely don't create it.

One misconception that some people have of the Institute for Advanced Study is that everyone there talks to everyone else. The Institute regulars find this rather amusing. "The Institute," says anthropologist Clifford Geertz, who's been there for sixteen years, "is not an intellectual club where I talk to Gödel about the incompleteness theorem and he talks to me about the religions of Java. That just isn't the way it is."

The way it is, by and large, is that people go off and work by themselves. "That's one of the features of life here," Freeman Dyson says. "We are very much separated into cliques, and we may regret that, but it's real, and that's how the work gets done. If I would make a point of going to sit down with a historian, or a mathematician in order to broaden my mind, I'd be neglecting my job in a way. My job first of all is to know the people in my business. So that's what I do."

Dudley Shapere, though, thought that this was a shame. "What really violates the spirit of the Institute," he says, "is the absence of interconnections among the various parts of it. The various schools are totally independent, and this is really a great weakness. Even the visitors rarely get anything from any other area. This is a personal gripe now, but with all the scientists at the Institute you'd think there would be a strong philosophy of science emphasis, but what do they have instead? Archeology, American history, that sort of thing. These are obviously important and good subjects in their own right, but there isn't the kind of cross-fertilization that ought to be going on at a place like that."

Shapere, anyway, made it a point to go outside his own discipline and view the progress of science firsthand. He teamed up with the one scientist at the Institute who is perhaps most influenced by Kuhn, namely, John Bahcall. Despite his Kuhnian doubts about the abstract theoretical possibility of doing so, there's one thing that Bahcall would love to know *The Truth* about, and that's solar neutrinos.

According to theory, neutrinos are produced inside the sun as the result of thermonuclear fusion. The sun fuses light elements into heavier ones, specifically hydrogen into helium. Helium is about four times as heavy as hydrogen and so it takes about four hydrogen nuclei to make one of helium. When the hydrogen nuclei fuse together, though, a slight amount of mass is lost in the process. Some of it is converted into energy as per the equation $E = mc^2$ and this energy radiates out in the form of visible light. The rest of it comes out in the form of neutrinos, in accordance with the so-called proton-proton sequence of reactions, in which four protons (hydrogen nuclei) fuse to form a helium nucleus, two positrons, and two neutrinos. Schematically, this is

$$4\text{H} \rightarrow \text{He} + 2e^{+} + 2\nu,$$

where H is a nucleus of hydrogen, He is a nucleus of helium, e^{+} a positron, and ν a neutrino.

Now this type of reaction is thought to be so well understood—it's the same process that underlies the thermonuclear reaction in the hydrogen bomb—that John Bahcall was convinced it ought to be possible to calculate the exact rate at which solar neutrinos should be falling to earth. If the incoming neutrinos were then observed to fall at the calculated rate, then our whole understanding of solar combustion would be confirmed.

So Bahcall did the calculation, inventing a new unit—the solar neutrino unit, or "SNU," specifically for the purpose. All that remained was to rig up some experiment that would measure the actual rate of the sun's neutrino bombardment of earth.

That was the hard part. Neutrinos are massless, which means that they travel at the speed of light; they're also electrically neutral, which means that they hardly interact with ordinary matter. It's thought, in fact, that the average neutrino could pass through *100 light years of lead* with only a 50-50 chance of its being absorbed. If neutrinos can do this, how in the world can any of them be expected to be stopped by a terrestrial-sized detector here on earth?

The answer is that neutrinos are thought to be raining down to earth in such vast quantities—billions passing invisibly through your retinas every single second of the day and night—that there's a high probability that at least *some* small number of them would interact with terrestrial matter if special measures were taken. This was a challenging problem for experimental physics, but Bahcall's colleague, Ray Davis, a chemist at the Brookhaven National Laboratory on Long Island, had an idea for designing a solar neutrino detector. His plan was to place a large tank of dry-cleaning fluid (perchloroethylene) about a mile below the earth's surface.

Standard perchloroethylene (C_2Cl_4) contains a large amount of chlorine-37, a chlorine isotope which—when it reacts with neutrinos—would be converted into argon-37. Each chlorine-37 nucleus is composed of 17 protons and 20 neutrons. If one of those neutrons absorbs a neutrino, that neutron will become a proton. As soon as this occurs, what had been chlorine-37 suddenly becomes something else, for its nucleus now has one extra proton (18 protons) and one less neutron (19 neutrons). In other words, it becomes an atom of argon-37, an isotope of the radioactive gas, argon. The upshot is that, if you find any argon-37 floating around in the vat of pure perchloroethylene, you know that either a neutrino or a stray cosmic ray has entered the tank.

It was precisely in order to guard against stray cosmic rays that the whole apparatus would be put so far below ground level. With money supplied by the National Science Foundation and by the Energy Research and Development Administration, Ray Davis deposited some 100,000 gallons of the dry cleaning fluid—enough to fill an olympic-size swimming pool—in a shaft of the Homestake Gold Mine, 4,500 feet below the village of Kellogg, South Dakota.

"If you go there," John Bahcall says, "there is a small hotel which is the only place in the world where I know you can get a neutrino martini. There are not many neutrino astronomers in the world—maybe three or four at best—but on the other hand we constitute a large fraction of the trade at that particular bar. It's also very close to where Calamity Jane and Wild Bill Hickok were assassinated, but for reasons which I presume have nothing to do with the experiment."

Anyway, they proceeded with their experiment. The only trouble was, very few, if any, neutrinos were ever detected by Ray Davis's apparatus.

"The first year that this result became clear I visited the mine with Davis," Bahcall says, "and we were very disappointed personally. We both had our careers on the line, and we had sold the government and lots of fellow scientists a very expensive project when a lot of people weren't all that enthusiastic about it. He wasn't getting anything, and I felt I was being proven wrong, and it was a disappointing summer.

"I can remember when we were upstairs in the room where you put on your boots and lots of gear in order to go down into the mine. We were in this room with the miners and we were putting on all of this gear, and we were a little bit glum. Davis had been spending most of the year going back and forth to the mine and had become quite friendly with some of the miners. They were asking him how things were going, and he was saying that things really weren't going all that well, that he hadn't gotten any neutrinos, and the experiment just wasn't doing what I said it was going to do. Altogether the situation wasn't too happy for either of us. There was a miner there who was very friendly, very pleasant, and he said to Davis, 'Never mind. Things will get better. It's been a very cloudy summer.'"

But things never got any better. Davis's neutrino detector was so sensitive that the experimenters could actually count, almost one by one, the number of argon-37 atoms in the tank—a mind-boggling feat given that these atoms would have to be extracted individually from out of 100,000 gallons of liquid. Their procedure was to bubble helium gas into the tank, stirring it through the liquid by giant paddles. Any argon-37 atoms present would be carried along by the helium. The helium would then be extracted from the tank, cooled, and passed through a charcoal filter where it would be separated from the argon. Because argon-37 is radioactive, its decays can be counted individually by an apparatus similar to a Geiger counter, and in this way the experimenters could keep track of exactly how many argon-37 atoms there were in the vat.

To check the whole process for accuracy, Davis and his colleagues added a known quantity of argon-37 atoms to the tank—exactly 500 of them. Twenty-two hours later they had gotten 95 percent of them back out again.

Based on his calculations, Bahcall expected Davis to be extracting 10 to 20 argon-37 atoms per measuring period, a span of a couple of months or so. To Bahcall's great dismay, Davis found many fewer, and in truth the number was not very different from what would be produced by random cosmic rays entering the tank. In other words, they could attribute little if any of the argon-37 to the elusive solar neutrino. "In fact," Bahcall says, "there is no conclusive evidence, after over ten years of the experiment, for any neutrinos at all being detected."

For Bahcall and Davis, this was unbelievable. There was no good explanation for it. Not that there were no explanations *at all*. If anything, there were too many of them. They fell into three main categories, having to do with the sun's production of neutrinos, the decay of neutrinos in flight, and problems with the detection process. But all of them were unacceptable for one reason or another, and some of them—like the idea that the sun is not in fact shining now—were on the bizarre side. (This is not quite as nutty as it sounds: the idea was that the sun generates its energy in spurts, and that we are now in a low period, so the neutrino infall rate is less than what it was supposed to be.) Another explanation had it that the sun's energy comes not from the proton-proton reaction, as had been thought, but from an unobserved black hole in the sun's interior. There was, of course, no independent evidence of any such black hole.

At the Institute, John Bahcall, together with visiting members Nicola Cabibbo and Amos Yahil, proposed their own explanation, which was of the "neutrinos-never-reach-us" variety. Neutrinos decay in flight, they speculated, and disintegrate into other particles. The only problem with this explanation was that there is no known particle for the neutrino to decay into. In the best tradition of particle physics, therefore, Bahcall and his colleagues invented a new particle expressly for the purpose, a low-mass scalar boson. Unfortunately, they could think of no test that would verify its existence.

So all the alternatives were uncongenial, some of them more unpalatable to the astrophysicists, some to the particle physicists.

"Almost every physicist I have met," John Bahcall says, "believes that the problem is in the astronomy. Almost every astronomer says that the problem must be in the physics."

The whole situation was so distressing to John Bahcall that he ended up comparing it to a Kuhnian paradigm shift: "We are now confronted in astronomy with a situation which is similar to that described in Kuhn's book on scientific revolutions. The reason is that there is this widely believed, widely used theory of how stars evolve and age, how they gain their energy and, in particular, why the sun shines. The fact is that the theory has failed the test, and this has led to people behaving more or less in the way Kuhn described in his book."

For Dudley Shapere, on the other hand, the failure of the solar neutrino experiment means only that *The Truth* is still out there, and it's only a matter of time before science finds what it is. The question will be settled by experiment, perhaps by the one that Bahcall, Davis, and others are now designing. It makes use of a germanium-gallium detector, which may cost upwards of 25 million dollars. If and when this experiment is performed, we may come to understand why it is that no solar neutrinos were ever observed a mile under the Black Hills of Dakota.

Humanists may argue among themselves over the epistemological and metaphysical status of science, but the scientists, even when—like

John Bahcall—they have philosophers' doubts about the final truth of their conclusions, seem to go on about their business as if all those issues were totally beside the point anyway. And if you should have the terribly poor fortune of being a philosopher yourself, you certainly don't want to hear what it is that the working scientists themselves have to say about you and your subject.

"I don't want to get into an argument with philosophers," Murray Gell-Mann once said. "I even have a prescription ordering me not to argue with philosophers. It was given to me by a doctor who attended one of my extension courses at ULCA."

"A lot of scientists are temperamentally allergic to philosophical discussions," John Bahcall explains. "I think it was Leibniz who described philosophy as the discipline in which you kick up a lot of dust and then complain you can't see. That's an attitude which many scientists share."

And why not? Philosophers are forever telling scientists what they can't do, what they can't say, what they can't know, and so on and so forth. In 1844 the philosopher August Comte said that if there was one thing man would never know it would be the composition of the distant stars and the planets. But three years after Comte died physicists discovered that an object's composition can be determined by its spectrum no matter how far off the object happens to be. But if the philosophers dictate to the scientists, the converse is not true: scientists spend comparatively little of their time handing down strictures to philosophers.

Of course it has happened. "Philosophers," claims Richard Feynman, "say a great deal about what is *absolutely necessary* for science, and it is always, so far as one can see, rather naive, and probably wrong. For example, some philosopher or other said it is fundamental to the scientific effort that if an experiment is performed in, say, Stockholm, and then the same experiment is done in, say, Quito, the *same results* must occur. That is quite false. It is not necessary that *science* do that; it may be a *fact of experience*, but it is not necessary. For example, if one of the experiments is to look out at the sky and see the aurora borealis in Stockholm, you do not see it in Quito.

"'But,' you say, 'that is something that has to do with the outside; can you close yourself up in a box in Stockholm and pull down the shade and get any difference?' Surely. If we take a pendulum on a universal joint, and pull it out and let it go, then the pendulum will swing almost in a plane, but not quite. Slowly the plane keeps changing in Stockholm, but not in Quito. The blinds are down too. The fact that this happened does not bring on the destruction of science."

Freeman Dyson was once asked about the philosophers who, like Paul Feyerabend and Thomas Kuhn, claim that science can't really get hold of the truth. Dyson told a story about the Albert Einstein 100th anniversary celebration which took place at the Institute for Advanced Study in 1979. The Institute formed a committee to invite speakers in three areas: science, history of science, and philosophy, and they were given lists of people whom they might consider inviting.

"The amusing thing was when we came to look through the lists," Dyson says. "We took the lists of scientists, and these were all people we knew personally—no problems. We took the list of historians of science, and these were people whose names we had heard of, even though we didn't know them. Then we took the list of philosophers of science, and these were people whose names we hadn't even heard of. I found that interesting. I mean there's a whole culture of philosophy of science out there somewhere with which we have no contacts at all. So when you talk about Feyerabend, well I haven't read a word of Feyerabend. I don't know anything about him. It happens just by accident that I know Tom Kuhn, but again I don't read much about what he writes. There's really little contact between what we call science and what these philosophers of science are doing—whatever that is.

"Very few scientists actually think as philosophers of science do anyhow," Dyson continues. "I mean most of the time we're dealing with real things, astrophysics, of course, particularly. I speak out of total ignorance because I don't read what they write, but I've listened to a fair number of talks by philosophically inclined people, and first of all they seem to consider—this is true of Tom Kuhn also—that quantum physics is essentially the whole of science. Somehow that is the one thing that they think is characteristic of science. Actually, quantum physics is probably very special; most of science isn't like that at all. Certainly not astrophysics."

The scientists at the Institute for Advanced Study search after the truth about things mostly without regard to what Shapere has spoken of as "the barriers to the possibilities of human thought" that have been erected by some of their humanist friends across the campus.

"It's ironic," says Dudley Shapere. "One of the major doctrines in philosophy of science in the past two decades has been that there are always many—even an infinite number—of alternative theories that can explain equally well any body of alleged facts, and that any one of these theories can be defended 'come what may.' This at the very time when physicists are finally coming to believe that it may be possible to get *one* theory, the theory of superstrings, that may explain everything! No wonder scientists think that philosophical ideas are irrelevant."

So Institute scientists proceed to think their vast thoughts about the extremes of nature, the immense and the infinitesimal, the distant past and the far-off future. Even the most philosophically minded and skeptical among them, John Bahcall, proceeds on a day-to-day basis as if he's sure that his work will tell him the truth about nature: not just a likely story or a provisional hypothesis or yet another "interpretation," but . . . *The Truth.*

To answer one simple question—Do neutrinos reach earth from the sun?—John Bahcall is willing—nay, eager; nay, anxious!—to undertake an experiment that will cost a mere 25 million dollars.

No indeed, a good scientist is not humble.

Life, the Universe, and Everything

Chapter 10

Nature's Own Software

When Stephen Wolfram came to the Institute at the advanced age of twenty-three, he was put into a first-floor corner office in the astrophysicists' building. Wolfram didn't really belong there, though, because he wasn't an astrophysicist. But he didn't belong with the particle physicists, either, because he also wasn't a particle physicist. Stephen Wolfram was in a new category altogether, one for which there was as yet no name.

Later, when the Institute gave him a whole suite of offices, to accommodate himself, his staff, and their combined computer gear, there was still no name for the type of physics they were doing, although for a while they thought of themselves as the dynamical systems group. The reason there was no name for what they were doing is that the field didn't exist yet: no one had ever done it before.

Most scientists restrict themselves to one narrow subject matter—to globular clusters, for example, or solar neutrinos, or fruit flies—but Wolfram had a far grander goal in view. He wanted to explain not the complexity of any given phenomenon, but *complexity itself,* wherever it might be found, whether in the structure of galaxies, or in turbulent fluids, or in the nucleotide sequences of a DNA molecule. He wanted to understand complexity, what's more, not in terms of the usual vehicle of mainstream physics, which is to say the differential equation, but in terms of something that was essentially new in science, the abstract, pattern-generating mechanisms known as cellular automata.

Cellular automata are not real things, they're only abstractions, creatures of the intellect. But they're big with Wolfram and his cohorts because it turns out that, when these imaginary mechanisms are simulated by a computer, they replicate the operations of physical systems that are actually found in nature. This is a bit uncanny. It's as if someone wrote a novel—an utter fiction—and then discovered that everything in the novel had actually happened.

There was the time Wolfram produced the seashell pattern, for example. He was working with a simple cellular automaton—the computer program for it was utterly innocuous, just a few lines long—and this diamond-shaped pattern shows up on the screen. It reminded him of some mollusk shells that he had once seen in a marine biology catalog. So he went back and paged through the catalog, and sure enough, there it was. He put the picture from the catalog next to the picture on the computer screen, and there was just no doubting it: his cellular automata simulation, the little self-repeating formula that he'd typed into the computer, it had produced exactly the pigmentation patterns found on the seashell. Hard to believe, but the proof was right there in front of him:

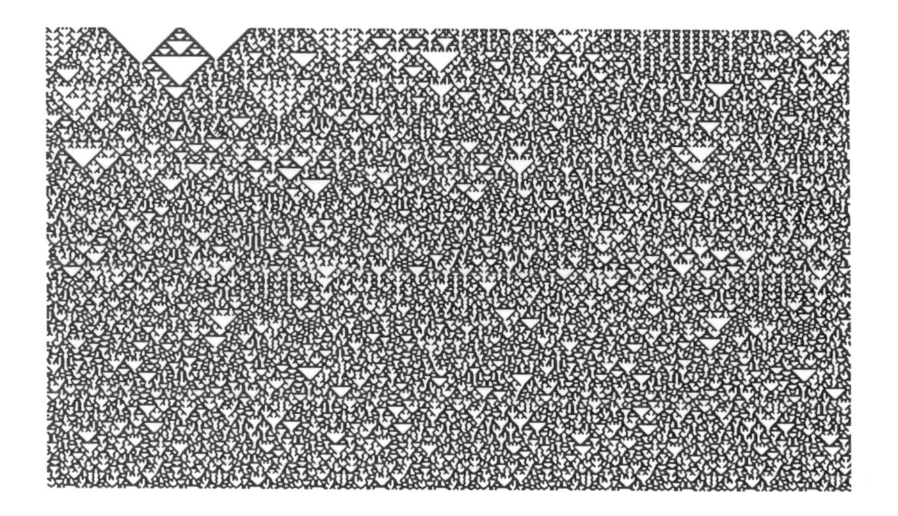

Later he discovered that cellular automata not only produced the patterns found on seashells, they also simulated the structure of snowflakes, the growth of crystals, the meandering of rivers, and any one of a dozen other things. It was incredible. In went a few lines of computer code, out came the real world, as if by magic. Cellular automata, Wolfram decided, might be able to explain the very architecture of nature.

Could it be, he wondered, that nature itself is in some sense a gigantic cellular automaton? Could it be, in other words, that the universe as a whole is a vast . . . *natural computer?*

Stephen Wolfram was born in 1959, in London, to a mother who was an Oxford philosophy professor and a father who was part-time import/export businessman and part-time novelist. Wolfram wasn't much interested in any of that, but it was clear right away that he had some special talents of his own and an unusual view of his place in the world. He went to Eton, where he played cricket—or at least he showed up on the cricket field. "I learned the best positions on the field so that I could read a book during the match," Wolfram says in his British accent. "I *had* to play cricket. I wouldn't have played it by choice." Later, at seventeen, he attended Oxford University—or at least he showed up on the campus. "I had the good fortune never to have to go to courses or anything like that, because I learned what I needed to know from books. From everything I've seen, courses are just a waste of time and one can learn most things a lot more quickly just by reading about them."

This isn't just idle boasting, because by the time he was fifteen, two years before enrolling at Oxford, Wolfram had written—and published—his first scientific paper, about a problem in particle physics. He never liked the article very much, though. "It wasn't very interesting," Wolfram says. "It's really lousy. I don't even have a copy of it anymore."

He has a copy of his second paper, though, which he wrote "much later," as he puts it, "when I was sixteen." He's saved a copy of this one, and he can even find it when challenged: "Neutral Weak Interactions in Particle Decays," published in *Nuclear Physics* in 1976.

In 1978, Wolfram came to the United States, to Caltech, where he was invited by physicist Murray Gell-Mann. He got his Ph.D. degree in theoretical physics a year later, when he had just turned twenty. "I actually missed out on being a teenage Ph.D.," he says, a little regretfully. Not long afterward, Wolfram received one of those MacArthur Foundation "genius" grants, the youngest person ever to be awarded one. You don't apply for these grants: the procedure is that you just get a phone call out of the blue one day to learn that you've won a tidy, tax-free sum every year for the next five years, and that with this money you can do whatever you want. Wolfram got $125,000; he chose to continue on with his research.

At the time, Wolfram's interests were divided between particle physics and cosmology. He was particularly interested in the evolution of the early universe, and he decided to work on the problem of galaxy formation. In order to do some calculations he found that it would be helpful to have a computer language that could handle algebraic expressions—abstract formulas—instead of just numbers. There was no computer language in existence that was really good at this, and so he decided to invent one of his own. Together with a few collaborators at Caltech—Chris Cole, Tim Shaw, and others—Wolfram created a language that would do algebra.

"Instead of just telling one that 2 + 3 is 5, for example, the program could tell one that $(x + 1)^2$ when expanded out is $(x^2 + 2x + 1)$. In other words, it could deal with symbols as well as with numbers." Wolfram named the language SMP, for Symbolic Manipulation Program.

It turned out that a symbol-juggling computer language had wide applications not only in theoretical physics but also in engineering and other branches of applied science. Wolfram saw no reason not to market the product commercially, and so he licensed SMP sales rights to a software firm called the Inference Corporation, of Los Angeles. This displeased Caltech, however, which claimed that it owned the language, since it had been developed on its premises and by its employees. Caltech and Wolfram settled their differences out of court, but Wolfram ended up quitting anyway in order to join the Institute for Advanced Study. At the Institute, they had a reputation for leaving you alone, something that was powerfully appealing to Stephen Wolfram.

It was during his dispute with Caltech that Wolfram first became interested in cellular automata theory. He was finding that, in order to derive the structure of galaxies from the fireball of the Big Bang, you needed some sort of pattern-generating mechanism. Cellular automata, it turns out, excel at the task of generating patterns.

"If you think about the thermodynamics of the early universe," Wolfram says, "you get into a strange problem. The universe is supposed to have started out as this uniform ball of hot gas, but in the end what we see is a lot of galaxies that are very patchy and irregular. The question is how do you get one from the other? Standard statistical mechanics says that you can't, and so I got interested in systems where you could start off with something which is completely random and completely uniform and end up with something which is kind of patchy and not uniform, and which might have some complicated structure to it."

At its most fundamental level, the problem involved goes back to the birth of philosophy, back at least to Plato. The question is how to get order out of disorder, complexity from simplicity. How from the chaos of the Big Bang do we wind up with structures as intricate as the chambered nautilus, the human eye or inner ear, the foundation of life itself in the bafflingly complex structures of DNA? It's the same problem that underlies the debate between science and creationism: You can't get something out of nothing, you can't get fantastically complex orderliness out of utter and irreducible chaos. To get what you actually find in the world, say the creationists, you have to suppose that God himself created it.

Wolfram, being a scientist, was intent on explaining order without reference to divine miracles. But if God didn't impose the order we find in nature, and if that order wasn't always there to begin with, then it follows that the universe must be somehow *self-organizing*. It must have created its *own* order. But how? What was the mechanism behind it?

Simultaneously with this, Wolfram was working on the wholly different question of how to get minds from machines. "From the other side I got interested in problems about artificial intelligence," Wolfram says. "I realized that if you want to make things really work in artificial intelligence it's no good just to have a computer with a single central processing unit, you have to have a computer that can process lots of information in parallel, and so I got sort of interested in what were the simplest parallel processing computers. So I was doing these two things—on the one hand trying to make a simple model for self-organizing systems, and on the other hand trying to understand simple models for parallel computers."

The one thing that both these problems had in common was that they required a way of getting complexity from simplicity: complex galactic structure from an original uniformity, and complex computing abilities from elementary components. So Wolfram took it upon himself to figure out how you could systematically generate complexity from simplicity.

He knew from the general concept of recursiveness in mathematics—the procedure of defining something in terms of simpler versions of itself—that complicated structures could arise from simple beginnings through the repeated iteration of one or more rules, as happens, for example, in the game called "Life."

Life was invented in 1970 by Cambridge University mathematician John Conway. The game is played out on a vast cellular space, a two-dimensional plane divided up into "cells," such as those on graph paper or on a checkerboard. Each cell has eight neighbors, four at right angles, and four more at the corners. Cells can be either *on* ("alive") or *off* ("dead"). If a cell is on, it's filled in with a marker of some type; if it's off, it's left blank.

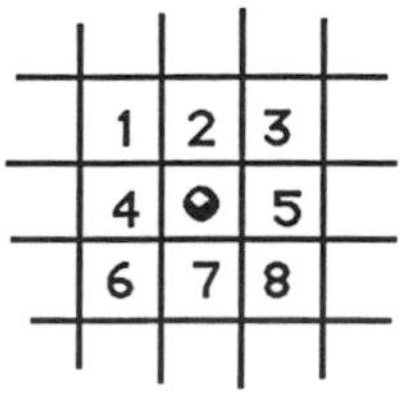

1 2 3
4 ● 5
6 7 8

The general principle behind the game is that life or death is a function of one's neighbors: isolated cells die of loneliness, while cells that are too crowded die of overpopulation. When neither of these extremes obtains, then live cells will remain alive, and when conditions are just exactly right then a live birth will occur. Just like in real life.

It all boils down to just two rules:

1. A live cell will remain alive in the next generation if it has either two or three live neighbors (the happy medium); otherwise it will die (from isolation or overcrowding).
2. A dead cell comes alive—a birth occurs—when it has exactly three live neighbors.

Those are all the rules.

Say, for example, that you start off with just two live cells, one right next to the other:

Now this is a sudden-death situation, because these poor fellows are too lonely to go on living. In the next generation both those cells will be off, and their squares will be blank.

But if you had begun with four live cells in a square array,

then everyone would be satisfied with life and they'd continue to live in the next generation, the reason being that having three live neighbors is the happy medium.

And if the blessed situation should obtain that a dead cell has exactly three live neighbors,

then a birth would occur:

Now one might think that from rules this simpleminded nothing interesting could possibly occur. But one would be wrong. Some starting patterns are like good genes: they are fruitful, and they multiply, sometimes in surprising ways. Take, for example, the T-shaped pattern called the "T tetromino":

In the very next generation (step 1 below), three births have occurred. In the generation after that, the shape breaks up, as if it were undergoing cell division, and then, births, deaths, . . . and ordered patterns emerge:

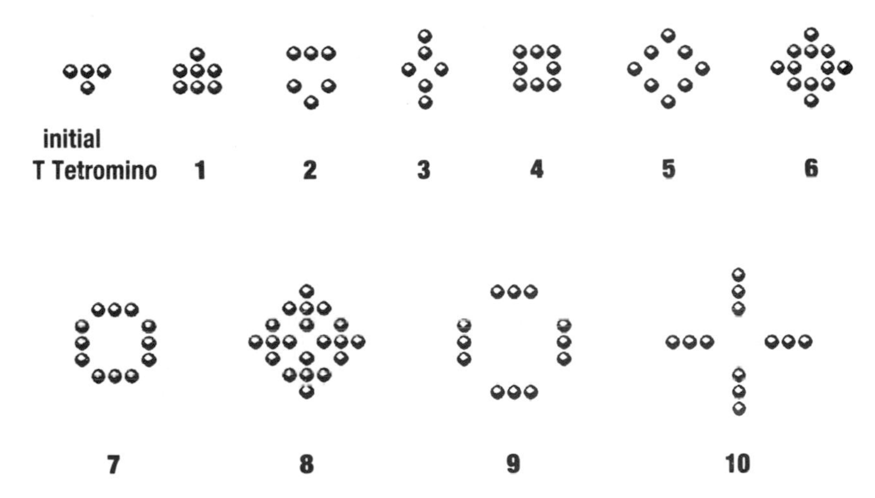

Turn the T tetromino clockwise 90 degrees, and add a single live cell on the upper righthand corner, and you've got an "R pentomino":

The R pentomino is incredibly prolific. After sixty generations (moves), it has exploded into a microcosmos (see next page).

Life's evolving patterns—"Life-forms"—are basic examples of cellular automata. They're *cellular* insofar as they exist in the squares—or cells—of a checkerboard-like grid. They're *automata* in the sense that they develop of their own accord—"automatically"—from repeated applications of the same two rules. In other words, Life-forms are not interactive: they require no human guidance or control for their growth and development. Given an

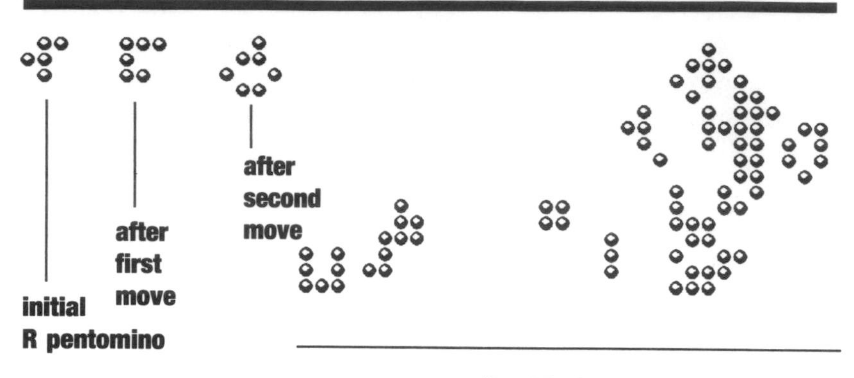

initial configuration of live cells, the entire future history of that Life-form is cast in stone for all eternity. The process is deterministic in the highest degree: from the same starting pattern you will always end up with the same results, no matter how many times you play the game, and no matter whether you make three moves or three million.

This, in fact, is what's so intriguing about cellular automata: they seem to grow spontaneously and unpredictably, but their behavior is as rule-governed and preordained as the most ironclad laws of physics.

The game of Life, originally set on a checkerboard, was tailor-made for the computer, and so hackers programmed it into their machines and watched the results. They found it so mesmerizing that, when Martin Gardner described it in his "Mathematical Games" column in *Scientific American*, professional computer users went on a Life binge. "My 1970 column on Conway's Life met with such an enthusiastic response among computer hackers around the world," Gardner said later, "that their mania for exploring Life-forms was estimated to have cost the nation millions of dollars in illicit computer time." For a while, Life addicts even published a journal devoted to the game, called *Lifeline*.

Hackers at M.I.T.'s Artificial Intelligence Laboratory—people like Bill Gosper, Ed Fredkin, and others—became so fascinated with it that they began to wonder if they weren't watching something which was more than just a game, whether they weren't in fact beholding some secret electronics-circuitry version of *real* life. The patterns on the computer screen seemed to act so realistically, the way they blossomed and died, spawned and merged—it was as if they were living bits of a digital soup. John Conway, the game's creator, took Life so seriously that he imagined Life-forms might actually *be* living entities. "It is no doubt true," he said, "that on a large enough scale Life would generate living configurations. Genuinely living, evolving, reproducing, squabbling over territory. On a large enough board, there's no doubt in my mind that this sort of thing would happen."

Ed Fredkin, for his part, allowed as how after all there was no way to definitively prove that *we ourselves*—living human beings—are not Life-forms in a game run by some metaphysical super-hacker.

Back in California, Stephen Wolfram had not yet met up with cellular automata theory, but he had heard about the computer game called Life. He was friends with Bill Gosper, who used to tell him all sorts of Life stories, and, although Wolfram was too levelheaded a chap to fall for Conway's and Fredkin's far-out scenarios about *truly living Life-forms*, he found the game quite amusing. After all, these strange Life entities were a little bit like the computer models that Wolfram was just then constructing on his own, to help him with the problem of galaxy formation. The main thing wrong with Life, so far as he was concerned, was that it had just one set of rules. What he needed was a way of studying the structures that would arise from many different types of such rules, and he needed a way of studying them all systematically, mathematically, not just as part of a computer game craze that was very big in hacker heaven, which is to say, at M.I.T.

"It was sort of amusing when I was thinking about these models of mine for a month or so," Wolfram says, "and then I happened to have dinner with some people from M.I.T., from the Lab for Computer Science, and I was saying that I got interested in these structures and I was telling everybody about them . . . and somebody said, 'Oh yeah, those things have been studied a bit in computer science, they're called cellular automata.' And I said that I've heard of those but that I don't know anything about them. Then I went off and looked them up and found all these papers and books about cellular automata."

Wolfram, though, was not impressed. "I was a little bit disappointed, actually. If you look up cellular automata on one of these computer searching things you'll find that there had been about a hundred papers written about them by 1981 or something, and so I went and looked up a whole bunch of these things, but they were boring. *They were so boring!* They were an illustration of a sad fact about science, which is that if someone comes up with an original idea, then there will be fifty papers following up on the most boring possible application of the idea, trying to improve on little pieces of details that are completely irrelevant."

Wolfram hates things that are boring. He's always done a lot of traveling, from one scientific conference to another, and one time he got the bright idea that he'd learn to fly. So while he was at the Institute he'd go down to Mercer County Airport, south of Princeton, and get into a Beechcraft Skipper, a small, one-engine training aircraft, and teach himself practical aerodynamics while his instructor more or less helplessly looked on from the right seat. That was exciting for a while . . . until the day Wolfram took off on a solo cross-country flight and got stuck in some faraway airport because the weather turned sour and he couldn't fly back. That was boring. And that was the end of Stephen Wolfram's piloting career.

Later, Wolfram traced cellular automata back to their origins in John von Neumann, who had showed how a vast arrangement of them could reproduce themselves in cellular space.

"Von Neumann had done something: he came up with the original idea," Wolfram says. "The idea was interesting, but the details, the construction he had made, it was completely boring. It's this book full of the design drawings of this completely weird object. The details of its implementation are like the most arcane mathematical proof one's ever seen. I don't know of any scientific thing that one learns from all those complicated details. I mean it's an interesting tour de force, it's an impressive proof—what he was trying to prove was that self-reproduction was possible, and he succeeded in proving that—but the method of proof was thoroughly arcane and complicated and I think not very illuminating as such."

But within a few months after his baptism into the world of cellular automata, Wolfram found himself at a scientific meeting on a small, privately owned island in the Caribbean. Here, in the shade of palm trees, and with the soft sea breezes wafting over him, Wolfram came face to face with a cellular automaton *machine*. This was not boring. This was interesting. "It was love at first sight" says Tom Toffoli, who watched Wolfram at the display screen.

The meeting had been arranged by Ed Fredkin, who, although he had never finished college, was on the M.I.T. faculty. Fredkin had made a fortune from a computer graphics and digital troubleshooting firm that he had founded called Information International Incorporated. By the time he sold it, Fredkin had made enough money to buy an entire Caribbean island. It was only a small spit of land, about three-quarters of a mile long and a half-mile wide—and called, quite appropriately "Moskito"—but it included a complete resort, Drake's Anchorage, with cabins and meeting rooms and one of the best restaurants in the whole British Virgin Islands. The place was, as M.I.T. grad student Norman Margolus said, "useful to own if you wanted to get people to come to a meeting."

The meeting, which was held in January, 1982, had grown out of an earlier conference on physics and computation theory held at M.I.T. the year before. It had been organized by the M.I.T. Information Mechanics Group, consisting of Fredkin, Norman Margolus, Tom Toffoli, and Gerard Vichniac. Scientists—not only computer scientists, but also physicists and mathematicians—were belatedly coming to realize that the computer was more than just a tool for doing number crunching, that in fact it seemed to mimic the world's processes in some hitherto inscrutable way.

About a dozen people came to Moskito Island, from M.I.T., IBM, the Argonne National Laboratory, and of course Caltech, from which came Stephen Wolfram. Sitting there at the computer watching the cellular automata march down the screen in waves of accretion, Wolfram began to realize their true possibilities. These things could produce whole ranges of textures, whole mini-universes. Sometimes, it's true, the patterns died out almost before they had started: not just any old set of initial conditions, apparently, could produce a universe. Other times the patterns started out chaotically but ended up producing a dull, repetitive order:

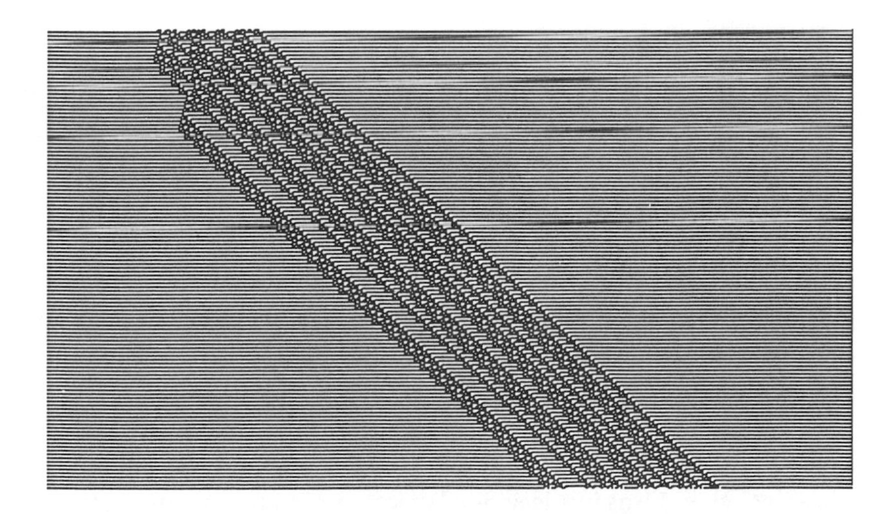

Still other times the patterns started out in an orderly way, and then degenerated systematically into nothingness, or almost:

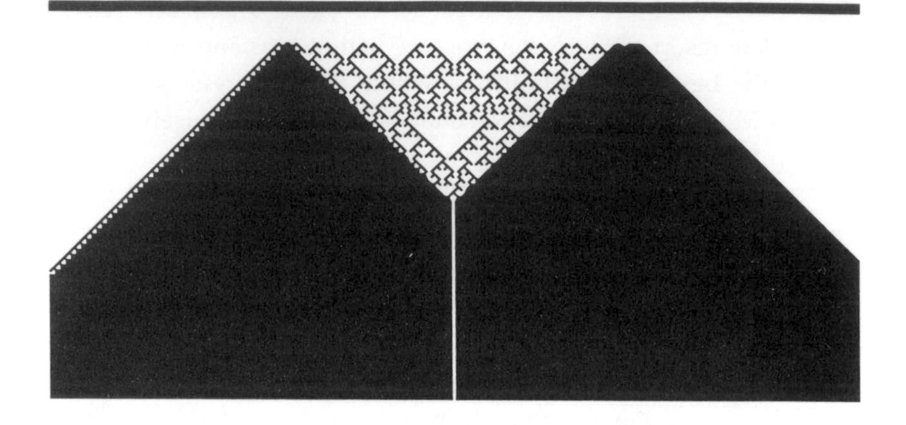

The fascinating part of it all was that there was no way to tell, prior to actually running the program, what the final result would be. To find out what happened, you had to get these cellular automata to work. They were mysterious, even a little eerie. You specified initial conditions, and you specified their rules of development, then you just cranked them into the computer and let it run, and a couple of seconds later—presto!—you had your own personal cosmos right there in front of you.

Some light was shed on the underlying principles by Charles Bennet, of IBM, who was also on Moskito Island for the conference. Wolfram talked to him about the theory of machine computation and cellular automata, and he came away with an even more expanded picture of how they worked and what they could do. Not only did they reproduce some of the physical structure of nature, they also seemed to illuminate the ways in which machines, and perhaps even people, processed information. It was as if cellular automata were suddenly the key to both matter and mind.

Wolfram was never quite the same afterward. "From that point on," Tom Toffoli says, "Wolfram's bibliography, his list of scientific production, goes from no cellular automata at all to 100 percent cellular automata. He decided that cellular automata can do anything. From that moment on Stephen Wolfram became the Saint Paul of cellular automata."

"That was in January '82," Wolfram says. "Between I guess February '82 and June '82 I got seriously started working on cellular automata. This was when I was just leaving Caltech, and in fact my two activities those months were speaking with lawyers, trying to un-mess-up this situation at Caltech, and working on cellular automata. In fact I think that probably for maximal personal happiness I should probably have spent more time at the lawyers and have made the science take a little bit longer and just done it a little bit less hard. As it was, I actually spent almost all of my time doing science."

A few months later, in December of 1982, Stephen Wolfram drove his red Volkswagen Rabbit to Princeton and set up shop in room 107 of the Institute's building E, which he turned into a cellular automata factory. He'd come in during the afternoon, program automata into his computers, and sit back and study his toy universes. Wolfram had three terminals in his office, two of them self-contained units, one of them connected up to the VAX in the basement. Sometimes all three would be running at once, sifting through patterns, searching through automata, trying to make some sense of them, to classify them according to complexity, longevity, and so on. Wolfram looked at hundreds, thousands, even tens of thousands of cellular automata patterns, often staying up late into the night, his computer display screens bathing the room with a dim blue glow.

Like the simple structures in the game of Life, every cellular automaton is a function of two things: an initial configuration of cells, and rules for producing new configurations out of previous ones. Wolfram talks about "sites" instead of "cells," and "time steps" instead of "moves," but otherwise the principles involved are generally the same as those in the game of Life.

"You take a line of sites," Wolfram says, "and each site has a value of zero or one, and the thing evolves in discrete time steps. After one time step you have a new line of sites. The value of a particular site depends on its own previous value, and the value of a couple of neighboring sites in the previous time step."

So you start with a single line of sites and then ask what's the value of the sites in the line below:

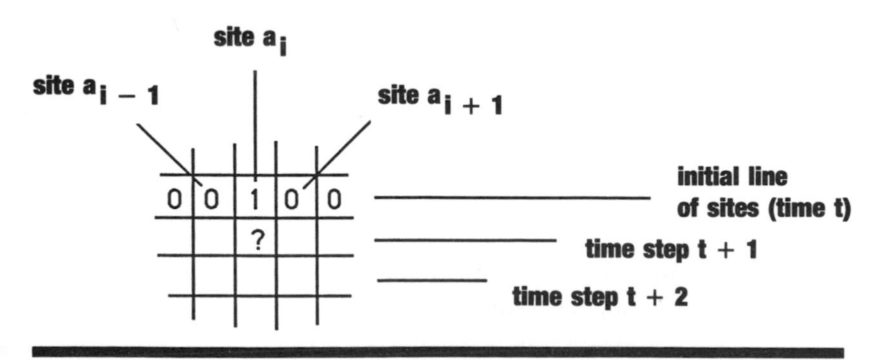

The value of a given site is determined by the values of sites in the previous line according to a rule. Wolfram states these rules mathematically, but most of them are much simpler than they look. One of the simplest cellular automata, for example, grows out of the rule:

$$a_i^{(t+1)} = [a_{i-1}{}^{(t)} + a_{i+1}{}^{(t)}] \bmod 2,$$

where

a_i is the initial site,
a_{i-1} is the site to its left,
a_{i+1} is the site to its right,
t is time, and
mod 2 indicates that the sum of the two site values is to be reduced modulo 2.

"All this means," Wolfram explains, "is that the value of each site is the sum modulo 2 of the values of its two nearest neighbors on the previous time step. In other words, the value of a given site at the next time step $(t + 1)$ is equal to the sum of the values at time t of a_{i-1}, which is the site to its left, and the value of a_{i+1}, which is the site to its right, reduced mod 2."

Mod 2. This refers to the sum of two numbers using modular arithmetic on the base 2. We already use modular arithmetic all the time, most of us without ever realizing it. "Clock arithmetic," for example, is addition mod 12: when you go into work at 9 o'clock and then put in 8 hours work, the resulting sum $(9 + 8)$ is not 17 o'clock, but 5 o'clock. In modular arithmetic, the only numbers allowed are those equal to or less than the base involved. Addition mode 2, therefore, works according to the table:

$0 + 0 = 0$
$1 + 0 = 1$
$1 + 1 = 0$

In plain English, what the mathematical rule above means is that, if sites a_{i-1} and a_{i+1} have different values, then the new site will have the value 1; if the two sites have the same value, then the new site will have the value 0. In this case, the value of the new site is 0:

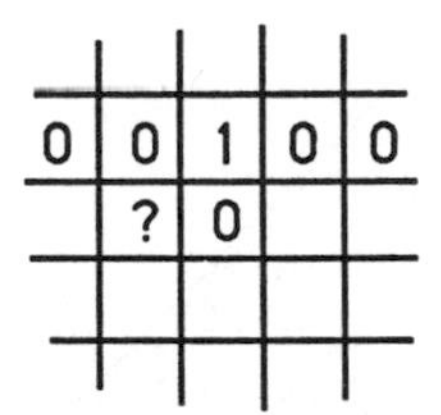

To find the value of the next new site, you apply exactly the same rule, applying it this time to the left- and right-hand neighbors of the site above it. You take the sum of those two values and reduce them mod 2. Since (0 + 1) mod 2 is 1, the value of the new site is 1. If you apply the same rule again and again to many new sites, a pattern begins to emerge:

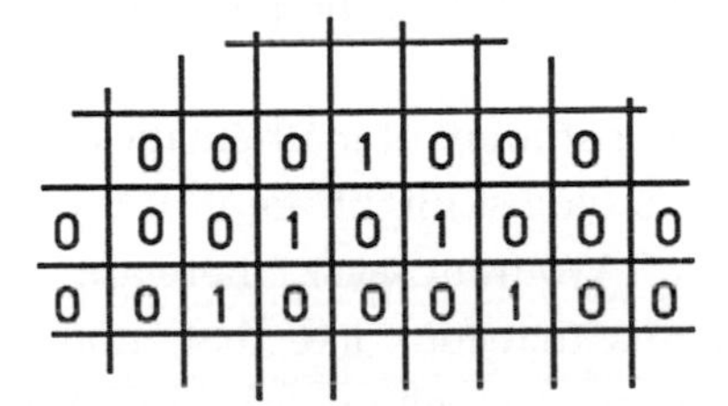

The automaton grows in this way through many additional time steps, ultimately producing a distinctly complex structure. After 23 time steps, it looks like Figure 8:

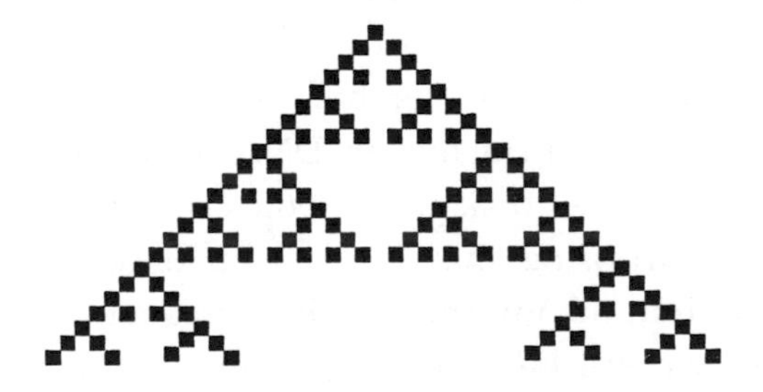

Figure 8 A cellular automaton pattern evolved from a single site according to the rule
$a_i^{(t+1)} = [a_{i-1}^{(t)} + a_{i+1}^{(t)}] \bmod 2$

And after about 100 time steps, it looks like this:

"What this shows," Wolfram says, "is that if you start off with just one nonzero site, the thing spontaneously grows into this kind of pattern. Even though the rule that formed it is very simple, the pattern it generates is relatively complicated."

The structure is recognizable to a student of mathematics as Pascal's triangle of binomial coefficients, but a biologist might see in it the pigmentation patterns of a snakeskin. "In fact," Wolfram says, "there are examples all over physics and biology of systems that look like that, that grow in exactly that way: crystal growth, for example, cell growth in embryos, the organization of cells in the brain, and so on. The important thing is that the mathematical features of cellular automata are the same mathematical features that are giving rise to complexity in a lot of the world's physical systems."

Cellular automata, or something like them, might be at work even in our own DNA.

"DNA is a very succinct program for how to build a creature," Wolfram says. "The number of bits in a human DNA molecule is equal to the number of bits on the larger disk drives you can buy today for computers, and it's awfully surprising that you can build something as complicated as a person just from the information that's in a medium-sized book. So clearly there's a very clever kind of programming that nature has done there, and in some sense what's going on in the development of cellular automata may be a bit like that.

"For example," Wolfram continues, "you might look at those seashell patterns and ask, 'How can such a complicated pattern be encoded in its DNA?' But if it's indeed the case—which one doesn't really know yet—

that those patterns are generated by the simple rules that generate cellular automata, well, then it's quite easy to see how DNA could encode such simple rules."

In the fall of 1984 Stephen Wolfram moved into new offices on the third floor of Fuld Hall. The Institute had created a suite of rooms for him and for the new staff he'd brought in, which included Norman Packard and Robert Shaw, two theoretical physicists from the University of California at Santa Cruz. Both had strong interests in complex systems theory and dynamical systems, which were now emerging as important new branches of physics.

Their aerie has the feel of an artist's loft, an impression created by the skylights, which have been built into the long, sloping roof overhead. They let in a subdued, diffuse luminescence that seems to come from nowhere, just the right type of light to prevent reflections in the computer screens. The loft is full of computers: there's an IBM AT, a Nova, a Ridge 32, plus three or four Sun Microsystems workstations. Wolfram is dependent upon computers for his cellular automata work, but all the same he doesn't much like programming. Programming is *boring*.

"I do an incredible amount of computer programming," he says, "but I don't particularly like it. In fact, just this weekend I wrote about a couple of thousand lines of code . . . and a lot of it was to do such wonderful things as to make my laser printer print out these pretty pictures. What took the longest time was getting the caption to come out right. Ugh."

Computers simulate the operations of cellular automata, but the weirdest possibility of all is that the reverse may be true: cellular automata may actually *be* computers.

"Some of these automata are quite strange and complicated," Wolfram says, "and I have a kind of strange speculation about them, which is that they could be used as universal computers. Given an appropriate initial state, perhaps you could encode in a program and data in such a way that the automaton itself would emulate the operations of a general digital computer. In other words, there would exist some initial state of the automaton that would make it behave like a computing machine."

Wolfram's idea draws back to the game of Life, to John Conway and Bill Gosper. After inventing the game and watching his Life-forms grow, Conway wondered whether there was any finite configuration of Life cells that would grow without limit, ballooning endlessly until it filled the entire

Life universe. He suspected there wasn't, and offered a $50 prize to anyone who could find a perpetually proliferating Life structure. Martin Gardner announced the challenge in his "Mathematical Games" for October 1970. Barely a month later, Bill Gosper collected the money. He had discovered the "glider gun."

The glider gun is a configuration of Life cells that spews out new cells on a regular basis. It's as if spontaneous creation were at work, for the new cells—"gliders"—just keep coming out and traveling away of their own accord, like bullets from a machine gun. A glider gun in good working order could transform a modest-sized initial configuration of cells into a teeming Life universe (see Figure 9).

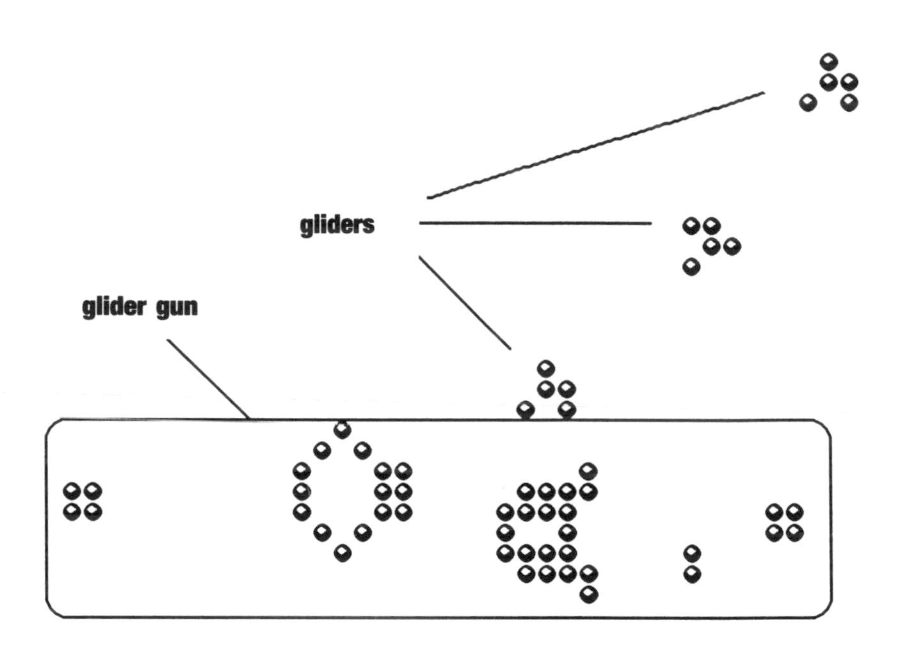

Figure 9 Bill Gosper's glider gun spewing out gliders

Because Life is a deterministic game that proceeds by deterministic rules, the emitted gliders behave predictably. If they meet no other Life-forms, they continue on forever. If they collide with another glider, then, depending on the angle of collision, they'll either annihilate each other or form other structures of known kinds, even including new gliders. This predictability, combined with the discrete character of Life-forms, led Conway to suggest that Life patterns could function as universal computers.

A digital computer, after all, is essentially a device that reduces all information to binary states—such as on and off, zero and one—and to simple functions such as AND, OR, and NOT. Conway realized that the right combination of glider guns and gliders could do everything that computers do. A stream of gliders and the spaces between them could stand for a succession of binary digits, ones and zeroes. Glider collisions could be logical gates, with the incoming glider stream being the input data, and the debris resulting from the collision being the output. If five gliders collide with four gliders and twenty gliders come out as the resulting debris, then you've got multiplication occurring. If ten collide with two and five come out, then you've got division. In principle, there was absolutely nothing that the correct combination of glider guns, gliders, and other Life-forms couldn't accomplish. Memory would be provided for by configurations of different Life-forms across a vast patchwork of cellular spaces. Conway even went so far as to prove, mathematically, that the game of Life was logically sufficient to encompass all the functions of a universal digital computer.

Although Conway proved that the game of Life was capable of universal computation in a theoretical sense, he never actually put together an array of Life-forms that worked like a computer. Cellular automata are more varied and powerful structures than Life-forms, and so Wolfram thought that they ought to be capable of universal computation. He therefore spent some time looking for the cellular automaton that would compute.

He wasn't successful, and so he followed in the footsteps of Conway. Conway had broadcast his challenge to the world through *Scientific American,* and so too would Stephen Wolfram. In the May 1985 issue, "Computer Recreations" columnist A.K. Dewdney wrote: "Wolfram suggests that lurking among [cellular automata] are true computers, vast linear arrays of cells blinking from state to state and churning out any calculation a three-dimensional computer is capable of. Wolfram, currently searching through the myriad of one-dimensional cellular automata is not above enlisting the help of amateurs in this sophisticated and daring enterprise."

It only took a month for Bill Gosper to solve Conway's problem, but to date nobody has ever solved Wolfram's, notwithstanding the flood of cellular automata structures that soon poured into his Fuld Hall offices. A cellular automata that computes, evidently, is a far more complex and elusive structure than a Life-form that merely proliferates *ad infinitum.*

Wolfram, nevertheless, is more than ever convinced that automata theory is important for the development of advanced computers, especially parallel processing machines. Conventional digital computers are built on the principle of "serial architecture," meaning that all operations are executed in the same sequence, performed one after the other. Because electrons stream through computer chips at almost the speed of light, serial architecture has been fast enough for most applications, but there's a fundamental way in which such a structure is at odds with the world at large, for nature works not serially, but in parallel. Out there in the real world, things happen all together and at once, as nature updates all its entities simultaneously from second to second.

The planets of the solar system, for example, affect one another through their gravitational fields, but they all exert their influence simultaneously. It isn't the case that Mercury's gravitational field influences Venus and *then* Venus's influences Earth's. Rather, all the planets affect each other in unison. For a conventional serial computer to simulate such planetary dynamics, it would have to do everything sequentially, calculating the first planet's effect upon the second, then the second upon the third, and so on. A parallel processing computer, by contrast, would more closely approximate what happens out there in the real world. If each planet were represented by a separate processor, then all processors could work at once and each could "feel" the gravitational influence of all the others simultaneously.

The parallel processing computer would not only reach its answers more quickly, its very principles of operation, its very structure, would approximate that of the external system it was attempting to emulate. If, for example, a computer had as many individual processors as there were cells in the living organism whose behavior it was trying to mimic, then each chip could stand for a single biological cell and the computer itself would in a sense "be" the whole biological organism. The computer's information processing would in that case reflect the structure, form, and evolution of the organism with a kind of fidelity that's impossible to achieve with ordinary, serial-architecture computers.

For Wolfram, the intriguing thing about cellular automata is that they have every sign of being natural abstract models *both* of physical phenomena in all their parallelism *and* of the parallel processing computers now being developed. It's as if cellular automata were the analog to both nature itself and the mind that beholds it. Both nature and computers are mechanisms that process information: nature proceeds from a set of initial conditions to a set of ultimate conditions, whereas computers proceed from input to output. Nature operates according to physical laws, whereas computers operate according to the instructions contained in their programs. But cellular automata are models of *both* these processes, the initial positions of a cellular automaton corresponding equally well to the initial conditions in nature *and* to the initial data in a computer. The evolution of automata, furthermore, corresponds to the operations of a computer *and* to the evolution of nature itself. And finally, the cellular automaton's rule of development corresponds to natural law on the one hand and to the computer's program on the other.

Nature itself, in other words, may be one vast computer, and mechanisms similar to cellular automata may contain its programming. Cellular automata, the software of the universe.

Stephen Wolfram's principal activity is abstract scientific research. "If I wanted to go out and make a bunch of money," he says in his Institute office, "I wouldn't be here doing research. I'm more interested in research than in making money."

Making money, though, is Wolfram's hobby. "Some people make furniture and sell it as their hobby," he says. "I develop practical applications of computer science and sell that."

First, there were the cellular automata postcards, six different automata in full living color, each described on the back of the card: "The color of each cell is determined by a simple mathematical rule from the colors of neighboring cells on the line above it." On the bottom of the card there's a statement of the mathematical rule that generated the pattern: "Rule 522809355 = 2032314344410_5," and then a copyright notice, "©1984 Stephen Wolfram." On the face is a picture of a cellular automaton, the colors corresponding to the values of each site. The cards span the whole gamut of the cellular automata universe, including the cellular snowflake.

Wolfram's colleague Norman Packard generated a cellular snowflake from a hexagonal rule ("Hexagonal rule $42 = 101010_2$," the postcard says), and the result, in blue, red, and purple against a black background, is surreal. When Wolfram wrote an article for *Scientific American* ("Computer Software in Science and Mathematics," in the September 1984 issue), the full-color snowflake was given almost a page by itself. In the same month, *Omni* magazine ran a story hailing Wolfram as "the new Einstein." A month later, in October of the same year, *Nature* featured seven full-color pictures of cellular automata on its cover, and, on the inside, ran an article by Wolfram entitled "Cellular Automata as Models of Complexity." Stephen Wolfram, it seemed, had arrived.

Wolfram had a circular printed up describing his postcards—as well as the much larger cellular automata posters that he was also thinking of selling—and he'd distribute the circulars at conferences, meetings, and the like. Some of the Institute regulars were a bit taken aback at the prospect of their otherworldly Platonic Heaven being turned into a mail-order computer graphics distribution warehouse, but Wolfram certainly didn't see it this way. He wasn't going to make any money selling postcards: they cost almost as much to make as he was selling them for. Selling postcards was just fun.

But they were only the beginning. There were the other cellular automata ideas, for wallpaper patterns, murals, and so on, the point here being not so much to make money as to bring cellular automata to the world.

"One of the things I've been meaning to do is to make a bit more of a serious effort to use cellular automata in some kind of computer art," Wolfram says. "In fact I have a little project which is going to happen sometime soon . . . well, there's this group at M.I.T. that's built a computer-controlled spray-painting machine. It produces 14 by 48-feet images—I guess it takes about twelve hours. It's in a warehouse around Cambridge, and in the next couple of months we're going to make some huge murals of computer art. Then there are some amusing ideas, such as creating a huge cellular automata display which one can put on the side of a building. Unfortunately, this turns out not to be technologically feasible for a feasible amount of money."

By the spring of 1986, word of cellular automata had gotten out of the closed circle of scientists and reached into the art world. "A month or so ago," Wolfram says, "I got this letter inviting me to an art exhibition in New York City, art based on cellular automata." It was at the Cash-Newhouse Gallery, in Greenwich Village. Wolfram went to the show. "It was kind of interesting, actually. I had expected something rather boring, but in fact the pictures were quite nice."

No matter how much freedom the Institute for Advanced Study gave its members, the fact remained that the Institute was not the best place in the world to be if you were at all commercially minded about science. Wolfram had some ideas about how to program a massive parallel processing computer, and so he worked for a time at Thinking Machines, of Cambridge, Massachusetts, which was developing a computer called "The Connection Machine." Largely the brainchild of Danny Hillis, the connection machine was planned ultimately to have one million processors linked up in parallel, making it more like a genuine biological brain than any computer yet invented. For a while Wolfram's scholarly papers gave two affiliations, the Institute for Advanced Study and the Thinking Machines Corporation, sometimes listing the latter of these first. Nothing exactly wrong with this, of course, but it was gradually becoming clear to Wolfram that he might be better off, might have a greater sense of freedom to do as he wished, if he left the Institute. It would be best, in fact, if he had an institute of his own.

In September, 1985, Wolfram drafted a two-page paper with the title, "Plans for an Institute for Complexity Research." The document described an institute staffed with a dozen senior scientists and another dozen post-doctoral fellows, plus technical and administrative support personnel, all of them devoted to fundamental research in the theory of complexity. As he envisioned it, the institute would engage in the commercialization of any ideas that had concrete applications.

"Fundamental research would be the primary purpose of the Institute for Complexity Research," Wolfram wrote. "Of greatest ultimate significance will probably be the development of very general principles of complexity. But in working toward such principles, it is essential to maintain contact with real applications that address existing problems. It will often be worthwhile to carry such applications through to the point of contact with practical technology. This Institute [Wolfram's *own* institute, not the Institute for Advanced Study] could sometimes, therefore, become involved with technological development, possibly in association with outside laboratories or corporations. Major new technologies might well lead to the formation of 'spin-off' companies."

The new institute, as Wolfram conceived it, would be attached to a major university, so as to have contact with faculty members in many other areas of science, as well as a supply of high-quality students. It would be funded either by a major donation, to be used as an endowment, or by grants from a university, corporations, government agencies, or any combination of these. And flowing from all this activity would be first-class scientific research, as well as major technological innovations that would return income to its supporters as well as to its staff. Wolfram had already done some work on an unbreakable cryptological system based on cellular automata that produce random patterns, and there was much more in the offing.

By the spring of 1986, Wolfram was in contact with some twenty universities that had an interest in housing his new institute. The only question was where to put it. "Some of the best opportunities, in terms of funding and so on," Wolfram said at the time, "are in the midwest, but who wants to go there? People will come to San Francisco or to Cambridge, they're nice places to live. But the midwest?"

In the fall of 1986, though, Wolfram quit the Institute for Advanced Study and moved to Champaign, Illinois. He now had *his own institute,* the Center for Complex Systems Research, at the University of Illinois. He brought with him his whole Princeton dynamical systems group, Norman Packard, Robert Shaw, Gerald Tesauro, as well as a few others, and today they're all housed in a modernistic low-slung brick building on the Illinois campus amid fifteen Sun workstations, assorted laser printers, and all the rest of the trappings of a small complex systems empire. They've also got a data link to a Cray XMP, one of the world's biggest and fastest supercomputers, which is housed next door, at the National Center for Supercomputing Applications. All things considered, Stephen Wolfram did not make out too badly.

"It's not as scenic here as at the Institute," he admits. "We only have a couple of trees out front, not a whole forest."

Not everyone, of course, thinks that cellular automata are the wave of the future. "Very pretty pictures," says Institute mathematician Deane Montgomery, rather dismissively.

"Wolfram has a dream that he's somehow or other going to understand complexity," says Freeman Dyson, "and that the complexity of the real world is mirrored in cellular automata. It's a big gamble."

The risk is that the connection between the patterns produced by cellular automata and those produced by nature is accidental rather than essential. It could all be a big coincidence. But Wolfram has seen too many parallels and deep connections between these abstract mechanisms and nature herself to think that all of it's going to evaporate away into a fog of sheer happenstance. The more he works, the deeper and wider those connections get—and Wolfram seems to do little else but work.

I last saw Stephen Wolfram at the Institute for Advanced Study one fine spring day in 1986. We were standing on the steps in front of Fuld Hall, looking out over the sprawling grassy acreage across which, in the old days, Toni Oppenheimer, Robert's young daughter, used to ride her horses. The sun was setting behind the trees, casting a deep orange glow on the high clouds, and Wolfram, as always, was telling me about his work, about his plans for cellular automata research, about the future of complex systems theory, about how he spends an average of thirteen days a month traveling to and from scientific conferences, and so on and so forth. As usual, I was bowled over, staggered, overwhelmed, by the man's intellect, and by his capacity for sheer work.

I used to wonder, so finally I had to ask.

"With all this constant traveling and work and science . . . uh; do you find the time for any social life? You know, girlfriends and such?"

"Oh yes," Wolfram said. "If you're interested in complex systems, there's nothing more complex than that."

Chapter 11

Beyond the Invisible

The rumors had been going around for a while, rumors that this might be *it*, the new theory of the universe, the theory of *everything*. Ed Witten was going to give a lecture at the Institute for Advanced Study, a talk in which he conceivably might outline the one and only true theory of nature.

That was the rumor. But even if it wasn't true, even if he wasn't going to present the theory of everything, Witten's talk would nonetheless be extraordinary, for at an institute where everyone is chosen simply and solely on the basis of their intelligence, on the quality of their minds, Edward Witten was known as being on a higher level altogether. Scientists aren't much given to saying things like "He's another Einstein"—that's journalists' hype—but no one had any trouble acknowledging that Witten was at the very least, well . . . incredibly smart.

It's Monday, November 12, 1984, and people are jammed into the lecture hall waiting to find out if the rumors are going to come true. Ed Witten is about to deliver the annual Marston Morse Memorial lecture, named after the late Institute for Advanced Study mathematician. The talk is held in the auditorium of the new library, right behind where Kurt Gödel's office used to be. Hugely modern in here, the room is equipped with electric blackboards, slate panels that float up and down at the touch of a button to give the speaker more writing space. There are lots of mathematicians in the audience, for the talk is sponsored by the Institute's School of Mathematics, but the whole contingent of Princeton particle physicists seems to be here, too, as well as a bunch of astronomers. Anyhow, the room is packed, with about two hundred of the best minds in science. They're waiting now, all of them, waiting to see if they're indeed going to be treated to the ultimate theory of the universe. You sure couldn't tell from the announced title, "Index theorems and superstrings."

At the appointed hour John Milnor introduces Ed Witten—not that he needs any introduction to *these* listeners—and then Witten launches into his talk. There's a strict allotment of time here and he doesn't want to go over, so he speaks very fast, in the high-pitched, quiet sort of voice that he has. You have to pay close attention to hear him, especially at this rate. The words and equations come pouring out of his mouth in a seamless stream, as if of their own accord, as if he could hardly stop this headlong rush of symbols and terminology even if he wanted to.

Witten speaks for a solid hour and then abruptly stops, saying, almost as an afterthought, that this is a new theory of the universe.

The people in the audience sit there like sphinxes, in great and stunned silence.

John Milnor asks if there are any questions . . .

But there are no questions. It's as if a wave has washed over the heads of these assembled intellects and now they're gasping for breath. They've just beheld the new vision of nature, the key to the cosmos, the secret of creation—it's been spread out in front of them in all its technical minutiae—but people's minds have gone blank, blotto, *tabula rasa*. There were a few recognizable phrases sprinkled in here and there—"tensor," "left-handed fermion," "Yang-Mills gauge groups," and so on, but most of the stuff that Witten was saying was so new to them, and so alien, that he might as well have been speaking Navajo.

"An introduction to superstrings should have taken fifty lectures, not just one," Ed Witten says later. "So naturally I wasn't able to explain more than a few aspects of it."

Almost a year later, on Thursday, October 10, 1985, Ed Witten gives another lecture at the Institute for Advanced Study. It's to be one of a series—not fifty, but anyway "a few," according to the bulletin-board flyers: "Professor E. Witten (Princeton University) will give a few lectures on Introduction to Simple Strings for Non-Physicists." From all those qualifying phrases in there you'd think that anyone would be able to follow these lectures: *Introduction* to *Simple* Strings for *Non-Physicists*. Even a child ought to be able to follow an *introduction* to *simple* strings for *non-physicists*.

The talk is held in the same room as a year ago, and Ed Witten is now walking back and forth in front of the big blackboards, waiting for the hall to fill up. He's tall and thin, with an Afro-scale ball of black hair dwarfing a long face and small pair of black horn-rimmed glasses. He's thirty-four years old.

Soon about a hundred people are packed into the lecture hall: the particle physicists, Adler, Dashen, Dyson, and their whole crew of post-docs; the astrophysicists, Bahcall, Don Schneider, and all the rest; and of course the mathematicians, Montgomery, Langlands, Bombieri, and so on. Also in the audience is Ed Witten's father, Louis Witten.

This is not merely a show of fatherly pride. Louis Witten is himself a theoretical physicist, specializing in gravity theory, and is a long-time member, trustee, and vice president of the once quite wild and woolly Gravity Research Foundation of Gloucester, Massachusetts. The Foundation was started back in the 1940s by stock-market mogul Roger Babson as a way of encouraging scientists to find ways of controlling the force of gravity. Babson thought that it ought to be possible to create a gravity shield or an antigravity machine or some other sort of device that would channel gravity toward purposes more constructive than merely making things fall, holding them to the ground, and so on. In 1949 the Foundation started to sponsor an annual essay competition, offering prizes of up to $1,000 for scientific papers advancing "some reasonable method of harnessing the power of gravity." Five years later, in 1954, two young Institute members, Stanley Deser and Richard Arnowitt, submitted a paper entitled "The New High-Energy Nuclear Particles and Gravitational Energy," which won first prize and got them the $1,000. Oppenheimer, then director, didn't think this was quite seemly behavior for his physics members to be engaging in, and he let them know about this in no uncertain terms. But at least he didn't make them return the money.

Later on the Foundation turned to fostering gravity research of the more traditional kind, and these days even the world's most respectable physicists have competed, including British astrophysicist Stephen Hawking, who won first prize in 1971. At any rate, Louis Witten has come to the Institute for Advanced Study this semester to learn the theory of superstrings, which, being a theory of *everything*, of course includes gravity within its scope. This is the theory that Witten's son will now try to explain to the assembled Institute members once again—this time expressly for any "non-physicists" who might happen to be in the audience.

The appointed hour—9:30 A.M.—rolls around and Ed Witten again starts talking in that tenuous voice of his, which sounds as if he's speaking from the Holland Tunnel or from inside the main ring of a particle accelerator. He talks completely without notes, pacing to and fro in front of the blackboards, putting equations up on them from memory. He asks rhetorical questions, puzzling ones, such as, "Why would a ten-dimensional world look like a four-dimensional world?," and then suggests answers, like "Nature started with something bigger than we observe and then subtracted something." But such remarks are mere bookends to the equations, where the real business of superstring theory gets transacted.

Witten writes equation after equation on the blackboard, quickly using up all the available space. The formulas refer to things like spinors, positive chirality, Dirac operators, and alpha prime. He raises one of the blackboard panels electrically and writes more equations, some forty-five of them by the time the hour is up. Gradually it becomes clear that "Introduction to Simple Strings for Non-Physicists" actually means "a talk that advanced theoretical mathematicians might have some chance of understanding."

When the hour is over, Ed Witten puts down the chalk, stops lecturing, and flashes a hesitant smile at the audience. They erupt in a round of applause, Louis Witten clapping away and beaming at his son the superstringer.

There's no moderator this time but—lo and behold!—a couple of questions have arisen from the audience. This time, evidently, people have come prepared. They had read some of Witten's papers and they had gotten some inkling as to what these "superstrings" were all about. They came knowing that superstrings were no more and no less than . . . the new building blocks of the universe.

Superstrings may portend one of the biggest upsets that the world of physics has ever seen. Ed Witten says, "We're still at the relatively early stages of a scientific revolution comparable only to the invention of quantum mechanics. It's just a vast process, one that's going to change everything we know in theoretical physics, at the really fundamental level. It's going to take decades. Maybe none of us will live to see it really come to fruition. But for a physicist, superstrings are life."

Superstrings, the new entitites of theoretical physics, are tiny one-dimensional curls of something or other (you don't ask what) that are on their way to replacing the old conception of elementary particles. The elementary particles—electrons and quarks and so on—used to be thought of as dimensionless points, as sizeless dots. They're dimensionless no longer, for now they have *one* dimension anyway, length. And they're dots no longer, for now they have a shape: they can be open strings, like a straight length of shoelace, or they can be closed strings, like a loop at the end of a lasso. Whatever their shape, these strings have the property that when they vibrate they behave like particles. Different vibration modes—frequencies, amplitudes, directions, and so on—are thought to produce the entire panoply of elementary particles. So this new theory says that at its finest, deepest levels of structure, the world is not made up of sizeless points, but of finite strings that bristle and throb so that they *look like* particles. If you could stop the vibrations, though, you'd see the underlying entities for what they really are. They're not "particles" at all: they're strings.

If all superstring theory did was to alter our notion of an elementary particle, it would already have dislodged a conception of atomic structure that goes back some seventy-five years. But superstring theory promises to do much more than this: it may in fact take us to what has been the Holy Grail of physics for the last half a century, to the Grand Unified Theory, the "Theory of Everything."

For the last fifty years, physicists have accepted two distinct theories as jointly covering all the fundamental laws of physics, quantum mechanics and general relativity. Quantum theory covers the small scale, relativity the large. The trouble has always been that, although the two theories are jointly exhaustive, meaning that together they encompass all of nature, they are also mutually exclusive, meaning that the one doesn't fit together with the other. It was quite mystifying. Nature is a consistent and self-agreeable whole from top to bottom—or at least physicists had always thought so—but the two theories that were the most successful in describing it somehow just couldn't be made to mesh together or match up. It was as if nature was having its private little cosmic joke.

But suddenly superstring theory looks as if it might change all that. Its underlying mathematical framework promises a unified picture of large and small, and if the theory comes through on its promise then the miraculous unification will be at hand. It will come as a gift, for free, like getting something for nothing. The sought-for unification will merely fall out of theory as a result of its sheer mathematical consistency. "Strings are a piece of physics of the twenty-first century that fell by chance into the twentieth century," Ed Witten has said. That's how magical it all seems.

The strangest part of superstring theory is that it's as much pure mathematics as it is physics. One thing that's for sure is that no one has ever *observed* a string, and it's almost as certain that no one ever will. Strings are just too tiny.

Elementary particles can be "seen" in the sense that their tracks show up in particle accelerators, gigantic machines such as those at Fermilab, in Batavia, Illinois, and at CERN, in Switzerland. By colliding particles into each other at very high energies, experimentalists can behold the tracks of entities as small as 10^{-13} centimeters across in size; anything smaller than that is currently invisible. Soon, with the coming of a new type of particle accelerator called the "superconducting supercollider," even tinier entities will be resolvable. But strings are even smaller, and by a wide margin. They exist on the scale of 10^{-33} centimeters. For experimentalists to be able to see them, their accelerators would have to be *light years* across in size. Strings are therefore worse than invisible: it's as if they're permanently and in principle hidden away at the finest interstices of nature, closed off from the prying eyes of empirically minded physicists. It's as if nature is saying, behind a turned hand, "Don't look, *think!* Damn the accelerators, trust the equations!"

Empirically minded physicists, of course, don't much like to hear that message.

"For the first time since the Dark Ages," say Harvard physicists Paul Ginsparg and Sheldon Glashow, "we can see how our noble search may end, with faith replacing science once again."

If "the superstring extremists" are right, says Alvaro De Rújula (who works at CERN), then "much of the fun of the discipline—to make empirically testable predictions and to bet on the results of experiments—would be forever gone."

But the experimentalist's loss is the theoretician's gain. Superstrings are in fact godsends for the particle physicists at the Institute for Advanced Study. No more messing around with accelerators and waiting on the edge of the chair for the results to be phoned in, it's now just a matter of getting all the equations right, of understanding the geometry involved, of keeping all the balls in the air at once. Superstrings are the Platonic Forms of physics. Abstract, remote, permanently unseeable, they're almost occult entities. Which may explain why Institute particle physicists look at superstring theory as if it would bring them to the right hand of God.

Ed Witten divides the history of superstring theory into three distinct periods which he calls the "Incredibly Primitive," the "Very Primitive," and the "Probably Still Primitive." The Incredibly Primitive era starts in 1968, with an Italian physicist's attempt to understand the "strong interaction." The most powerful force in nature, the strong interaction is what holds like-charged particles together against the electrical repulsions that would otherwise push them apart. The protons inside the nucleus, for example, are all positively charged, and, since like electrical charges repel, protons ought to be scattering from each other in a wild explosion. But the protons are in fact stuck together by a nuclear glue, the stuff of the strong force.

Although its workings were not well understood in the 1960s, the strong force was known to operate over a whole class of elementary particles collectively termed "hadrons," of which there were then more than a hundred known examples. Hadrons had the property that, when their angular momentum was plotted against the square of their mass, then the resulting curves formed relatively straight lines. These curves were called "Regge trajectories," after Institute for Advanced Study physicist Tullio Regge, who discovered them.

In 1968, Italian physicist Gabriele Veneziano, then at the Weizmann Institute of Science in Israel (and later a visiting member of the Institute), published an article that gave a mathematical treatment of the Regge trajectories. Veneziano had worked out a group of equations that could be used to generate the straight-line Regge curves. This was interesting in itself, but more interesting still was what it led to, for a year later, other theorists, beginning with Yoichiro Nambu of the University of Chicago, noticed that Veneziano's equations were consistent with a new conception of elementary particles. Instead of dimensionless points, particles could be viewed as having definite spatial extensions. They could be thought of as lines, as "strings."

According to the Veneziano-Nambu model, these strings were finely tuned balances of opposite forces: of tensions pulling the string ends together opposed by accelerations keeping the string ends apart. The string was in motion, like an airplane propeller, and the so-called centrifugal forces pulling the string ends apart were balanced out evenly by the string's inherent inward tension. This inward pull was enormous, some 13 *tons* per string. It was as if you could hang the weight of a half dozen Cadillacs from just one of these strings, and the string would hold . . . which, if true, would be quite a remarkable achievement in view of the fact that each string was conceived of as having zero thickness.

The idea of portraying subatomic particles as if they were strings was utterly new to theoretical physics. There were many technical (not to mention historical, philosophical, and aesthetic) reasons why particle theorists had always represented elementary particles as dimensionless points, so this new conception that particles were extended in space was unexpected, to say the least. Nevertheless, the string model had a couple of things going for it. One was that it made sense out of the Regge curves. Another was that the string hypothesis provided an acceptable model of "quark confinement," which is to say that it explained why experimentalists had never seen quarks in their accelerators, only the bigger particles made out of them.

Of an atom's three particles—electrons, protons, and neutrons—only one, the electron, is thought to be truly "elementary," meaning that it's not composed of anything more basic. The other two particles are thought to be composed of more primary entitites, quarks. A proton, for example, is supposed to consist of three quarks. One of the major problems with the quark model, however, is that individual quarks had never been seen in the laboratory. Try as they might, experimentalists just could not produce free quarks in their particle accelerators, a fact that led theoreticians to conjecture that individual quarks must be permanently imprisoned ("confined") somehow inside the larger particles.

But how? This was the problem of quark confinement, and theorists presented various different models as possible solutions. There was, for example, the "bag" model, according to which quarks were trapped inside containers from which there was no escape.

The string model was an alternative to the bag idea. From the string viewpoint, quarks were attached to the ends of strings. The reason why you never saw an individual quark all by itself was that one quark is tied to another by means of the string, so that they only came in pairs or—if you had a three-ended string (a slight infelicity in the string model)—in triplets. Try to pluck a quark from its string and you'd only succeed in stretching the string, or snapping it in two (or three), leading to the creation of additional quark pairs (or triplets) at the new string ends.

As a model for quark confinement, and as a way of explaining the Regge trajectories, the string model was all very ingenious, and it attracted a lot of attention, at least for a while. "Thousands of people worked on string theory back in those days, a lot of them in Europe," says the Institute's Roger Dashen. "You go back in the literature, you'll find several thousand papers on string theory. They weren't called strings then, they were called Veneziano amplitudes, dual-resonance theories, things like that. In the end, though, physicists realized that none of this had anything to do with reality."

It had nothing to do with reality because it turned out that in its ground state—meaning at its lowest energy level—a string was not merely massless, it in fact had a *negative mass*. What this could possibly mean from a philosophical viewpoint was an intriguing problem all by itself, but the physics of it, at least, was clear: particles with negative mass (called "tachyons") would travel faster than light. This was fully consistent with general relativity, which maintains that only particles *having mass* cannot exceed light speed, but it had rather unwelcome consequences, nonetheless, such as a backward flow of time, violations of the law of cause and effect, and others.

In addition, there was the problem of extra dimensions, for it turned out that the original string theory would make sense only if it were assumed that space-time had twenty-six dimensions instead of the usual four (three space and one time dimension). Even for theoretical physicists, who regularly entertained the most outlandish hypotheses with the most cordial tolerance imaginable, it was somewhat uncomforting to contemplate a 26-dimensional universe. ("Obviously unphysical," one of them remarked.)

As if that weren't enough, the theory of quantum chromodynamics (QCD) soon swept onto the scene as a complete and successful theory of the strong interaction, accomplishing exactly what string theory had been created to do in the first place. Strings suddenly seemed to have no reason for being—in fact they created more problems than they solved—and so most physicists pronounced them essentially dead.

Except for a few diehards—like John Schwarz, for example. Schwarz had worked on string theory almost from its inception, and he thought that the whole framework was aesthetically pleasing and that—even with all its flaws—string theory might perhaps be fashioned into something that could work.

Schwarz's background was right for a string theorist. He'd been an undergraduate mathematics major at Harvard, then changed to theoretical physics, getting his Ph.D. from the University of California at Berkeley in 1966. It was while he was teaching at Princeton that he first heard about Veneziano amplitudes, dual-resonance theories, and so on, and he had worked on virtually nothing else ever since. Schwarz, with the help of many others in the field, added two important elements to the "Incredibly Primitive" string model of Veneziano and Nambu. One of them was the concept of supersymmetry, the other was the idea of compactification.

Supersymmetry is the notion that the particles of matter—quarks, electrons, and so on—are mirrored in a deep and systematic way by corresponding particles of force. Every particle of matter is said to correspond to a "superpartner" particle of force, and vice versa. Whereas the original string theory of Veneziano and Nambu applied only to the so-called bosons (force-carrying particles), Schwarz, together with Pierre Ramond, André Neveu, Joël Scherk, and others, brought "fermions" (particles of matter) into the picture as well. The result was that a theory which had started out as an explanation of the strong force now seemed to encompass all the known particles.

Of course there were a few problems. For one thing, there were still a few extra dimensions lying about. Schwarz and his colleagues had, it's true, reduced the number of required dimensions by more than half. Their string model had "only" nine dimensions. Nine was not three, but it was a lot better than twenty-six. "A step in the right direction," Schwarz said.

Still, what are you going to do with six leftover dimensions?

The answer is that you roll them up into a little ball and sweep them under the rug. In 1974, Schwarz and Joël Scherk realized that the six superfluous dimensions could be eliminated by a mathematical device called "compactification," a technique pioneered back in the 1920s by Theodor Kaluza and Oskar Klein. Kaluza and Klein had attempted to unify electromagnetism with gravity by beginning with a five-dimensional spacetime manifold. Klein proposed that if the extra, fifth dimension were small enough, as compared with the other spatial dimensions, then it would not show up in macroscopic nature (at least it would not show up *as* an extra dimension, although it might manifest itself in some other form).

In essence, the process of compactification amounts to getting rid of unwanted dimensions by making them too small to be noticeable. A garden hose, for example, is an ordinary three-dimensional object having length, width, and height. But if you imagined the hose's diameter shrinking down almost to zero—as would seem to happen if you viewed a stretched-out hose from a distance—then the hose would *look* like a one-dimensional object. Two dimensions would have been "lost" by having been made too small to see. Nevertheless, they'd still exist, and would be perfectly real.

Schwarz and Scherk now applied the Kaluza-Klein compactification technique to the leftover six dimensions of their string model. They reasoned that if the extra dimensions were compactified into little spheres, too tiny to be seen on larger scales, then the whole string theory would begin to make sense.

Except for the problem of gravity. The most evident and obvious force in nature, the one that we're all familiar with practically from the moment we're born, gravity was still not comprehended under any theory of quantum phenomena. Quantum mechanics could accommodate the other three forces of nature quite handily. The electromagnetic force, for example, was completely embraced by the theory of quantum electrodynamics (QED), while the strong force was taken care of by quantum chromodynamics (QCD). The weak force was included as part of the electroweak unification, but as for gravity, well . . . it had stubbornly resisted any and all attempts to be understood quantum-mechanically and was still the odd man out.

The main reason for this is that gravity is described by Einstein's general theory of relativity as a curvature of space, a large-scale phenomenon in which massive bodies cause distortions of the surrounding space-time manifold. The problem with these huge yawning chasms of space, however, is that they're not obviously compatible with what goes on at the quantum scale. The disparity between the two extremes is just too great. This was the problem of "quantizing gravity," and it manifested itself through the presence of what particle physicists trippingly rolled off their tongues as "nonrenormalizable infinities."

The infinities in question arose from the very fact that the elementary particles are regarded as *sizeless*, as dimensionless points. A particle carries a force, and this force gets stronger the closer you get to the particle. The force reaches crisis proportions when you get zero distance away from a zero-sized particle, for in that case the particle's force field becomes infinite. Thus, "the problem of infinities."

The usual way around the problem was "renormalization," a corrective process which amounts to subtracting one infinity from another, leaving you at the end with a value of finite size. This second infinity could, for example, be considered to be the particle's mass when it is compressed down to zero radius. Subtract an infinite mass from an infinite force and you'll end up with a particle that has the finite mass it is supposed to have.

The infinities involved in quantum gravity were, however, special. They were not renormalizable. In the case of gravitation, the infinities are too big and arise too rapidly, and the subtraction process just does not work. Here, too, the crux of the problem seemed to be that the elementary particles of gravity—"gravitons"—are conceived to be pointlike. In string theory, on the other hand, particles are no longer points: they have dimensions, they have size and shape. If anything could avoid the dreaded nonrenormalizable infinities, chances were good that strings could do so.

Some versions of string theory predicted the existence of an unknown particle of spin-2. For a while, this was thought to be just another one of the weird and "unphysical" consequences of string theory, for no such particle had ever been observed, but in 1974 Schwarz and Scherk theorized that this unobserved spin-2 particle could be the quantum of gravitation, the "graviton," which physicists had determined would be of spin-2.

Schwarz and Scherk realized that if the spin-2 particle predicted by string theory were in fact the graviton, then string theory would have the near-miraculous result of *requiring* that gravitons existed, of *necessitating* them as an inescapable and ineradicable part of the theory, whereas all other quantum theories could not be made to accept gravitons through any amount of forcing, juggling, or mathematical hocus-pocus. Infinities would be prevented from arising in string theory because of the fact that strings had size. And so it seemed that the new string theory of Schwarz and Scherk would solve both the problem of quantizing gravity and the problem of infinities in one fell swoop.

During the Incredibly Primitive era of superstring theory, Ed Witten was just getting into science. He was born in Baltimore in 1951, attended the Park School in that city, and then Johns Hopkins, where he majored at first in history. Later he transferred to Brandeis University, in Massachusetts, where he decided to change his major to physics. He had a lot of science to catch up with after he changed majors, and so he doesn't remember his Brandeis experience too fondly. He came to Princeton for graduate work in physics, and worked under David Gross for his Ph.D., which he received in 1976 at the age of twenty-five.

The year before he got his degree, Witten attended a summer school in the Swiss Alps, near Mt. Blanc, where he met an Italian physicist by the name of Chiara Nappi, whom he ultimately married. Both of them landed positions at Harvard, later at the Institute, and finally at Princeton University.

In 1981, Witten was not yet working in superstrings, but he proved a result that would turn out to have an important application for string theory. The result was that, for technical reasons, *eleven* dimensions of space and time seemed to be a magic number. They were the minimum necessary to get a realistic point-field theory, but according to Witten they were also the maximum permissible. There was one major problem with eleven-dimensional unified theories, however, and this was that no such theory in an odd number of dimensions would be "chiral."

Chirality is the physicist's term for handedness, the fact that nature often distinguishes between right- and left-hand (or clockwise and counter-clockwise) versions of the same thing. People are chiral in the sense that their hearts are found on the left side of their bodies. DNA is chiral in the sense that the double helix twists around like a spiral staircase always to the right, never to the left. But the phenomenon of chirality also applies to elementary particles. Neutrinos, for example, are left-handed, meaning that they spin counterclockwise, or to the left, with respect to their direction of forward motion. In order to accommodate the chirality of the quantum universe, therefore, a unified field theory must be able to distinguish between right- and left-handed particles. One of the things that Ed Witten and others showed, however, was that, in order for a theory to be chiral, it must presuppose an *even* number of dimensions.

So matters were at an impasse. Witten had demonstrated two things that appeared to be fundamentally incompatible: that a unified point-field theory required an *odd* number of dimensions, *and* that a theory could not be chiral unless it were in an *even* number of dimensions. There seemed to be something inherently impossible about a unified theory that was simultaneously chiral and also based on point-particles.

This was bad news indeed for point-particle theorists, but not, of course, for string theorists. Strings are not point-particles, they're *strings*, and so they skirt Witten's dilemma quite neatly. The Schwarz-Scherk string theory was in ten dimensions *and* it included chirality as well, and for once the way seemed clear for the triumph of string theory. For once strings seemed to have everything going for them. They incorporated gravity without getting into the problem of infinities, they incorporated all the forces and all the particles, and they did it all without being subject to the embarrassing odd-even-dimension problem posed by Witten.

Could you ask for anything more?

Why, yes. You could ask for freedom from anomalies.

Pity the poor superstring! What Gottlob Frege had long ago said about the underpinnings of mathematics seemed to be perfectly applicable here: "just as the building was completed, the foundation collapsed."

"Anomalies"—the word conjures up freakish physical phenomena, monstrous and unexpected turns of events, surreal miscarriages of physical law. If Halley's Comet were to turn up again after only thirty years instead of its regular seventy-six, that would be an anomaly; two-headed cows, Siamese twins, these are "anomalies." But in particle physics the word means something less weird although much more serious from the point of view of abstract theory. In particle physics a theory is said to suffer from anomalies if it violates certain conservation laws, such as the laws of conservation of electric charge, the conservation of energy-momentum, and so on. Because these conservation laws are so firmly entrenched as fundamental rules of physics, any theory that violates them is regarded as basically an incoherent theory, and is automatically dismissed.

For a long time one of the major problems with string theory was the fact that *every* known version suffered from anomalies. It was as if they were built into superstrings, as if the theory itself were permanently jinxed. But in the summer of 1984, John Schwarz, along with Michael Green who had gotten interested in superstrings when he was an Institute member in the late 1970s, came up with a result that would finally validate string theory and set the world of physics on its ear. The two men showed that there were versions of superstring theory that were *anomaly free*.

It was like a sign from heaven, a revelation. Physicists finally began to pay attention to superstring theory. And why not? Here was a theory that not only incorporated gravity, it mathematically *required* it. What's more, purely by virtue of the one-dimensional, finite-sized structure of the string itself, superstrings avoided the infinities to which all other quantum theories of gravity had been subject. The fact that string theories were now proven anomaly free was like the pot of gold at the end of the rainbow, and physicists reached for the gold with their hands stretched out.

"I dropped everything I was doing, including several books I was working on, and started learning everything I could about superstring theory," said Nobel prizewinner Steven Weinberg in January, 1985.

"I think this may be it," said Nobel prizewinner Murray Gell-Mann. "The theory of everything—gravity, weak, strong, and electromagnetic interactions plus a lot of other things all together—a completely unified theory of nature."

"Superstring theory is a miracle through and through," said Ed Witten.

Major stories about strings appeared in the usual science magazines like *Nature*, and *Science*, but also in *Time*, and in the *New York Times*. Scientific articles on superstring theory began to pour forth at the rate of some one hundred a month, and some of the principals in the string wars, Princeton's David Gross, for example, were personally receiving an average of fifteen preprints a week. The July 1985 issue of the trade journal *Physics Today* carried a news story headlined "Anomaly cancellation launches superstring bandwagon." It claimed that, "In the last few months it's been difficult to find a particle-theory preprint that doesn't begin with: 'Considerable interest in 10-dimensional unified string theories has been sparked by the recent discovery of Green and Schwarz that such theories are free of anomalies." It went on to report that the world of theoretical physics was gearing up for a major and protracted assault on superstrings.

The general sense of euphoria got to be so heady, so almost out of control, that those who managed to keep a sense of perspective about string theory found themselves criticizing their brethren for overzealousness. Sheldon Glashow and Paul Ginsparg, for example, wrote for *Physics Today* a short satirical piece—cutely titled "Desperately Seeking Superstrings?"—in which they urged a bit of moderation. "Years of intense effort by dozens of the best and the brightest have yielded not one verifiable prediction [from superstring theory]," they said. "The Theory depends for its existence upon magical coincidences, miraculous cancellations, and relations among seemingly unrelated (and possibly undiscovered) fields of mathematics. Are these properties reasons to accept the reality of superstrings?"

Later, Glashow would go around to conferences quoting his witty Witten couplet:

Please heed our advice that you too are not smitten;
The book is not finished, the last word is not Witten.

Harvard's Howard Georgi, meanwhile, spoke of string theory as "recreational mathematical theology."

But true superstringers paid no attention. The wave of the future had crested, and there was no fighting it. Superstrings were here for good.

By the time Ed Witten was giving his short course of lectures—*Introduction* to *Simple* Strings for *Non-Physicists*—the Institute for Advanced Study had turned into a hotbed of young stringers. Those who weren't working on strings, people like Steve Adler and Freeman Dyson, felt they had to apologize, explain, or otherwise account for the fact.

"I tend to be a bit of a loner, and to work a little out of the mainstream," Stephen Adler says, "just sort of because that's sort of the way I like to work."

"I'm not as quick as the young people," says Freeman Dyson. "And it wouldn't be sensible for me to be part of that, so I tend to go off into the less fashionable things, like the origin of life, as an example."

The young post-docs, by contrast, were firmly on the superstring bandwagon, but like most everyone else in particle physics, they found it tough sledding. "It's a very tough field," says Mark Mueller, "and these are very, very complicated theories and it takes a long time to develop a facility with them."

Some of these stringers were even a bit concerned about their theory's lack of connection with empirical reality. The Institute's John Bagger, who was lucky enough to have Ed Witten as his dissertation advisor, explains, "Physics made great progress when Galileo said 'Open your eyes and do experiments. Stop doing philosophy. Start using your senses and see what nature's telling you rather than just think about it.' That's when physics became physics and that's why physics has been tremendously successful. It's very dangerous just to think about things and not do experiments. So it is a danger that strings might not be decideable at the high energies."

That said, the question is how to get around the problem. "Well, there *are* ways around it," Bagger says. "First of all, we don't really even know enough about strings yet as a formal theory to be able to tell if they're testable at low energies. But it certainly bothers me. I think it's an unhealthy situation, but at the same time strings are so *appealing!* . . . We were at an impasse in certain questions in quantum field theory and this is a way out, and so one has to explore it mathematically before one can actually propose tests for experimentalists."

Despite the dangers of the field, Bagger is going full speed ahead with superstrings because they may answer some of the Very Biggest Questions in physics.

"Superstring theory may tell us," Bagger says, "why the universe looks the way it looks. Why is it so large? Why isn't the whole universe the size of a marble or something like that? In other words, how does a theory know what's called the 'vacuum structure'? Forget about the planets, forget about people and all that. *Why is the universe as big as it is?* Nobody really knows the answer to that. This is the old question of the cosmological constant. Why is the cosmological constant so small?"

The cosmological constant is a parameter in physics that describes the structure of space in the absence of any massive object whose gravity would deform it. In other words, it describes the curvature of a space empty of all matter. The cosmological constant has been measured—in fact it's the single most accurate measurement ever made in the history of science, accurate to within one part in 10^{120}. It turns out that the cosmological constant is almost exactly zero, which means that empty space is almost absolutely flat.

"But *why* is it so close to zero?" John Bagger asks. "The answer is that nobody really knows why. The cosmological constant is basically the energy density of empty space, and since energy density is what makes curvature occur in relativity, if there's a big cosmological constant, the universe is very small, very curved up. And so the fact that the universe is so big is a mystery. *Why is the cosmological constant so small?*"

If superstring theory can solve that mystery, there are still other mysteries in the theory itself. Why is it, for example, that the visible universe has only four large spatial dimensions, while the other six are "compactified," hidden away into little balls? The superstring answer is that at the time of the Big Bang all ten spatial dimensions were more or less equivalent, but then some withered away almost to nothingness, leaving the four that now remain. This is what Ed Witten meant when he said that nature started with something bigger than we observe and then subtracted something. But *why* did some get small?

"How did nature know which of these spaces superstrings prefer?" Bagger continues. "That's really the key question for string theory: *How do you know that four dimensions are big and that six dimensions are small?* That's not known. I think physicists would like to believe that the universe is unique: it *had* to be the way it is, that God didn't have any choice in the way he made the universe. At the moment there is no real understanding of that. If there were a million possible worlds and God just chose one at random, then I think physics would be a lot less appealing because one is trying—basically with sheer thought, but also by interaction with experiment—to try and understand *why* the universe is the way it is. But if it's just an arbitrary choice on God's part, then the whole procedure's a little bankrupt."

Ed Witten's office is on the third floor of Princeton University's Jadwyn Hall. It's a small, sparse room that overlooks an open courtyard in the center of the building. The bookshelves are stocked with physics texts, and there are several file cabinets filled with Witten's papers in physics. There's also a computer terminal in the room, not that Witten uses it much for physics.

"Before you can put something into a computer," Ed Witten says, "you have to have a reasonable and rather specific understanding of it. Our understanding of string theory is still too general for that."

Witten uses the computer for writing, however. Together with Michael Green and John Schwarz, Witten is writing a book called *Superstring Theory*, to be published by Cambridge University Press in 1987. At least until the advance of theory outstrips the book's contents, this will be the definitive treatment of the subject. How long the advances are likely to take is anyone's guess. "Some of my colleagues think it's going to happen faster than I think," Witten says. "I think that anybody who believes this will happen within the next few years has underestimated the majesty of the framework. It may take a very long time, perhaps fifty years, as quantum electrodynamics did."

Witten, in a gray sweater, is eating a sandwich.

One of the problems with string theory, Witten says, is that it was invented backwards: the mathematics preceded the conceptual understanding, just the reverse of the way it was with Einstein. Einstein had the concepts first, then developed the theory.

"General relativity is a theory where the concepts came first," Witten says, "and since Einstein everyone is trying to imitate it. Einstein is really the first person who put the concepts first, instead of playing around with equations, like his predecessors Newton and Maxwell did. Superstrings, unfortunately, is more a mathematical instead of a conceptual theory. Strings were invented by chance rather than by being deduced from any logical framework. The problem with superstrings is that we don't understand it conceptually yet. One of the things I'm trying to do is to provide the conceptual framework that's still missing."

Witten is convinced that superstrings may finally be *The Truth* about nature, the one and only truth about the universe.

"A year ago I would have listed three seemingly insuperable problems standing in the way of the superstring theories: first getting the right low-energy gauge interaction; second the chirality problem; and third, explaining why the cosmological constant vanishes." The first two problems were removed by Schwarz and Green in August of 1984 and, Witten says, "made it completely clear to me that superstring theory is correct."

Witten is ready to work a lifetime, if that's what it takes.

"Superstrings are either the theory of nature," he says, "or they're an incredible step forward toward the theory of nature. And since a physicist studies nature and wants to understand it, trying to undertand this theory is going to be our careers, for as far into the future as one can imagine."

It's a fabulous time to be an Institute for Advanced Study particle theorist, almost as fabulous as it was in the legendary old days, back in the heyday of quantum mechanics in the early decades of this century. Some physicists used to wish they were alive back then, so that they could have shared in the excitement of witnessing those great leaps forward in the understanding of nature. But superstring theory has given physicists another chance to get in on the ground floor, to see the merging of the top-down and bottom-up theories of the universe.

You have to have a certain hubris to think that the final Grand Unification is ever going to happen. You have to have a certain extreme immodesty about the powers of the human mind.

"Theoretical physicists are very proud of what we've accomplished in quantum theory up to now," says superstringer David Gross, "but it's been based on data forced down our throats by our experimentalist friends. Now we're in a situation where there are no experimental clues, just a feeling that the theory must be right. Being where we are now—working at this threshold that's eighteen orders of magnitude beyond anything that we can test—takes a lot of arrogance, and there are physicists around who think it's not only arrogant but foolish."

It may, of course, be that the experimentalist is the foolish one. He spends his days, weeks, months, perhaps even years, waiting deep down in a mine shaft, waiting for a proton to decay or for a solar neutrino to light up his instrument panel. For a theorist, this is no way to live. Better to have a beautiful theory than to be twiddling your thumbs down in some salt mine waiting for a light to go on.

The eye of the theorist is ever on the Forms—the mathematical realities—whether they be at the bottom of matter, or at the edge of the universe. Abstractions, equations, pure mathematical consistency may in the end provide the full and final truth about nature. You may come to know the world using your mind alone, just like Plato always said you could.

Epilogue

Chapter 12

Babes in Toyland

When I was at Princeton in the 1940s I could see what happened to those great minds at the Institute for Advanced Study, who had been specially selected for their tremendous brains and were now given this opportunity to sit in this lovely house by the woods there, with no classes to teach, with no obligations whatsoever. These poor bastards could now sit and think clearly all by themselves, OK? So they don't get an idea for a while: They have every opportunity to do something, and they're not getting any ideas. I believe that in a situation like this a kind of guilt or depression worms inside of you, and you begin to worry about not getting any ideas. And nothing happens. Still no ideas come.

Nothing happens because there's not enough real *activity and challenge: You're not in contact with the experimental guys. You don't have to think how to answer questions from the students. Nothing!*

—Richard P. Feynman
"Surely You're Joking, Mr. Feynman!"

The old guys show up first. They're in there every day promptly at 12 noon on the dot. They open the place up. They're well dressed, in jackets and ties, just like in the old days. They bunch together, these small knots of elderly men, and they sit and eat their lunch in the Institute dining hall and they discuss . . . whatever it is they discuss. Old times, probably. For some of them, this seems to be what they come in for, to go to lunch.

They don't linger, though. Most times it's only 12:30 before they're back out in the open air again, walking to their offices, as if there's pressing business at hand, things that have to be seen to immediately. So these small, shrunken, wizened old men shuffle off to their old and forgotten book-lined rooms, where nobody comes to visit them, where nobody comes to ask for their sage advice, for the wisdom of their years . . . and they crack open a book and read it for a while. Soon enough, most of them are gone for the day.

But one or two of them, the real troopers, they stay on at the Institute all the way until tea time, which is to say 3 o'clock. The oldest of them, those who were writing their most famous books way back in the 1930s, before the Institute was even founded, they rise from their desks at about this time and saunter along to the common room where they pour themselves a cup of tea. They pick up a napkin and a cookie—and perhaps a newspaper from the rack there—and settle into a comfortable sidechair. Slowly they begin to nod off, dozing for just a moment or two—no more than that, certainly!—and after a while their chests are heaving slowly and deeply in the golden light of late afternoon.

Later, when they carry their empty cups back to the tea wagon over by the grandfather clock, their hands shake so much that you can hear the cups rattling in the saucers all the way across the room.

"So what's come out of the Institute, anyway?" Jerry Ostriker wants to know. He's sitting in the Greenhouse room of the Casa Lupita Mexican restaurant in Lawrenceville, New Jersey, just down the road from Princeton. This is the dinner that the astronomers are throwing for Margaret Geller after her lectures at the Institute and later at the university, where Ostriker teaches astrophysics.

"I mean," he continues, "there've been all these great names there, Einstein and Gödel and so on, but as far as the science goes, what has the place really produced?"

"Well, for one thing," someone says, "there's the Yang and Lee parity nonconservation work. And then of course there was Gödel's work on the continuum hypothesis. And . . ."

And that was about the end of it. The impression is hard to shake that the Institute is a nice place to visit but that you wouldn't want to live there, because . . . not enough really happens. The people there just don't *produce*. The place is so staggeringly free of pressure, its members are left so utterly alone to do what they want, that some of them end up not doing much of anything. "Back in 1950 when I was a junior fellow at Harvard," says a former member, "I used to say I'd love a research job. But people said, 'Oh you don't want that. Look at Professor X. He went to the Institute and hasn't done anything since.'"

Of course there are no rules about this: some people work as hard at the Institute as they would anywhere else. Still, the place seems to have something of a special aura about it, a dark lining, a lingering reputation that this is where the most promising people roll in like tidal waves and then . . . then they disappear meekly into the shallows of science.

Institute members themselves often wonder if their carefree existence really has been for the best. Freeman Dyson, over the years, has considered moving to a university, "to earn an honest living," as he puts it. "Several times I almost did. In the end what persuaded me to stay is the family—I raised six kids here. The kids didn't want to be uprooted. The only question is whether such an easy and pampered life is really good for one. That's hard for me to judge."

Particle physicist Steve Adler feels much the same: "I can't decide, in terms of the impact on my own research, whether it's better to have the forced sort of immersion in things produced by teaching, or it's better to just read up on the things you're interested in and have more time for your own work. I don't know the answer. You can't run your life twice. I don't regret having come here: it probably fits my style of working pretty well."

But if the Institute is not more successful than any other academic institution in getting its faculty to produce original, high-quality work, if its own members can't decide if the place is a boon or a bust, then what in the world is its reason for being? Why does it exist at all? Is it no more and no less than a motel, an intellectual resort where burnt-out scholars go for a vacation . . . or to reap their final reward for having been geniuses in their youth? Is the Institute the place where shy and retiring theorists go to eat intellectual candy and play with their intellectual toys?

The younger generation tends to show up a bit later for lunch. They've been spending half the morning arguing in the hallways, after all, and these arguments can go on forever. They're dressed casually, in jeans or corduroys, and often enough in T-shirts, perhaps a "theme" shirt, like the one on the astrophysicist there. It has a galaxy on the front, with an arrow pointing to one edge, captioned: "You are here."

When these youngsters finally do get to the dining hall, they, much like their elders, bunch together at the tables. They spend their lunch hour talking about their respective subjects—heterotic string theory, complex systems, or whatever—and they hang on in the dining room for quite a length of time before heading out again.

Back in their offices they spend the rest of the afternoon in front of the computer, debugging a program, or working on an article. If they do any reading at all it's likely to be a matter of checking a reference in a scientific journal before citing it in a footnote. Perhaps they return a few phone calls or reply to computer mail. At tea time the young people stand in small groups and laugh and joke, or—as is Stephen Wolfram's custom—they stop by long enough to pour some tea into a plastic cup and take some cookies back to their office, where they continue on with their work. Most of them stay at the Institute until 5 o'clock or so; some remain late into the night.

The Institute's younger members, at any rate, spend little time asleep. Many of them are at a critical point in their careers and they're at the Institute on a make-it-or-break-it basis. They're between jobs, perhaps, or trying for tenure at their home university, and the one thing they need most in life is to get several important pieces of research out of the door and published. For this specific purpose the Institute for Advanced Study is a godsend. Unlimited free time—no classes, no hassles—these post-docs know how to make the best of it. They don't need any fires lit under them.

Things are a little different for the permanent faculty. "Everyone who gets to the Institute as a permanent professor," John Bahcall says, "gets here because he's done—in all cases—two important things. Otherwise you don't get here."

The question is, of course, whether after having done two important things, you can do anything further. "It's very hard to do more than two important things in the sciences," Bahcall continues. "Or even to do *one*. So often your best work is done before you get here permanently."

But the Institute for Advanced Study is not at all unique in this, for the same thing is true of people who get the best positions anywhere: by the time they make it to the top, they're on the point of being finished as scientists. What, for example, do Nobel laureates produce *after* they've won their prizes? In many cases, nothing much. It's like being Miss America: you're the most famous person in the world for a while, and then you're never heard from again. Oh, everyone knows where the top people are—they're at Harvard or Caltech or the University of Texas or wherever, and they've got flocks of very bright graduate students under them—but as for important original research, well . . .

But of course there *is* something that's special and unique about the Institute for Advanced Study, and Clifford Geertz, of the School of Social Science, puts his finger on it: "Oppenheimer once said, so I heard, that the point of this place is to make no excuses for not doing something, for not doing good work. And in fact there *aren't* any excuses, so if you aren't doing good work everyone knows it, and the anxiety level is extraordinarily high. And so I think that nobody is well advised to come here who can't take that kind of pressure."

"No duties, only opportunities," Abraham Flexner used to say.

And then there's the Einstein mystique. The icons are all over the place. You can't miss them. There's an Einstein poster in the receptionist's office, there's a bust of Einstein in the dining area, there's a photograph of Einstein in the director's office, and so on and so forth. "I mean you're supposed to be a genius if you come here," Geertz says. "You're supposed to walk on water. Einstein did walk on water, but the rest of us don't . . . and then if you come here, and not only don't you walk on water but you don't even sort of wade, you're in trouble. Psychologically, that's a hard life to live. Here there is nothing to do except work, and if you don't do it, it might leave you a little nervous."

Lately, some members have taken to saying that the Institute has a particular problem in the School of Natural Sciences, that the quality of its physics faculty has declined gradually over the years until now it's at an all-time ebb. Younger members, especially, are prone to making this complaint.

"The Institute is on what I would characterize as a mediocrity spiral," the Young Turk says. It's late in the day and most of the Institute regulars have gone home by now, but the Young Turk closes the door to his office anyway. He's been a member of the School of Natural Sciences for a couple of years, and in fact he's on the verge of leaving the place, but there's no point in burning your bridges.

"This is true mainly in physics," he says, "but I gather that it's true also in the arts side of it. I don't know about that because I don't know a single person who works in history or social science, but to be honest, the people who are in physics right now—the senior group—are mediocre, at best. I mean they're not much good, I can tell you that. When a place has good people—you know, Einsteins, von Neumanns, and so on—then they hire other good people. But if its people aren't the best in the world, then there are two ways that it can go. Either they can always hire those who are worse than them, or they can be more selfless and hire people who are better than them. And the second thing is very rare. The usual group dynamic is that you only hire those who don't pose a threat to the existing group."

"As for example?"

"One rather well-known example is Ed Witten," the Young Turk says. "They talked about hiring him for a long time—this was before he was quite so famous, but it was clear that he was very good—and I really think that what it came down to was that various people just sort of felt threatened by him. It was as simple as that."

The Institute regulars, of course, will tell you that hiring decisions are never simple. "Being threatened is not the only reason for appointing bad faculty," one of them says. "Bad judgment is another. The opposite can happen, too. Mediocre people can appoint those much better than themselves."

Part of it also is the small size of the Institute's faculty, which magnifies the importance of any available position. "The jobs at the Institute are so precious," says Murray Gell-Mann, "they're all big jobs like main professorships at universities, chairs in England or something like that, that people get very nervous about whom they appoint, in physics at least. And that doesn't lead to very good choices. You don't get as good people as you would get if you were more relaxed about it."

At times, though, it does appear that a person can be too good, or at any rate too famous, for the Institute for Advanced Study. When, quite recently, there were two faculty vacancies in the School of Mathematics, a senior professor was asked whether the Institute would make an overture to Benoit Mandelbrot, the inventor of fractal geometry.

"I don't know that we'd want him," the mathematics professor says. "I mean, he's big in the popular mind, but is he really that big? I know what it says in the magazines. I'm sure he's a good man."

"Don't you think he's put forward a fundamental new idea in mathematics?"

"I'm not sure," the professor says.

"Why not?"

"You see these nice pictures that you get by simple patterns and so on. Very nice. I agree that there's something to study."

"Isn't it true that euclidean geometry doesn't capture the shapes of what we see in nature, the shapes of trees, clouds, and so on, whereas fractal geometry does that? And that therefore it's a key to nature in a way that no previous geometry has been?"

"I'm not sure about that, no," the professor says, shaking his head. "No. I know that you can read that anyplace you want to. It says that in all the magazines. But I don't know. I mean I've asked some people who know more about it than I do, and they don't seem terribly sure. I know that you can find plenty of people who would make academic appointments on the basis of what it says on the front page of the *Times* magazine section."

"But what is there *against* the fractal theory? Is there an obstacle or difficulty with the idea?"

"What would be a positive sign?" the mathematician says. "Is it really such a staggeringly new idea?"

"Well, you're the mathematician."

"And I'm telling you that it's not clear to me that it is."

"Have you studied Mandelbrot's books and papers?"

"No," says the professor. "But has anybody else who talks about him?"

The Institute did not, at any rate, hire Benoit Mandelbrot.

But even when the Institute finds someone who meets all its criteria, it may still not come up with the right combination of money and perks that will snare the man or woman in question. For Murray Gell-Mann, whom the Institute has tried to hire again and again, salary has always been a drawback.

"The reasons why I didn't go there," Gell-Mann says, "were very complex. They partly had to do with money, and with summer vacations, and things like that. My family and I were very attached to spending our summers in Aspen, and it wasn't so easy to see how the summer pay that we might get would cover going to Aspen. And if one didn't include summer pay, then the salary wasn't competitive."

As for Ed Witten, the Institute offered him a long-term membership when he was just getting his Ph.D. from Princeton, but Witten turned it down in favor of a research position at Harvard. Later, when the Institute had another chance at Witten, it failed to make a move on the grounds that since he hadn't been there even for a semester, the permanent faculty members didn't know him personally, and they didn't know how well he'd fit in. Princeton University, on the other hand, whose faculty members knew him quite well, because he had gotten his degree there, they quite happily took him into their ranks. Another opportunity missed, another chance frittered away.

Nor is Ed Witten by any means the only example. "In my opinion," Norman Packard once said, "it looks like the Institute has a certain conservative trend that may prevent it from being really first-rate. This may partly be a reflection of a personal prejudice. It's not clear, for instance, that the Institute is going to continue supporting the kind of research that Wolfram and Rob Shaw and I are doing, and obviously I think this is a really important research direction. If the Institute lets this whole research enterprise evaporate, then it will go elsewhere, and so will the vitality and life that this research represents."

Packard was speaking in January of 1986. Nine months later, Stephen Wolfram and Norman Packard and Rob Shaw and Gerald Tesauro—all of the Institute's complex systems people, in fact—had moved to Wolfram's new Center for Complex Systems Research at the University of Illinois.

To get the best people, the Institute either will have to cough up the inflated salaries of the good old days ("Back in Flexner's time," Harold Cherniss says, "money would not be an issue"), or it will have to start putting itself out on limbs. "Some people think that it's really not the purpose of the Institute to take risks like this, that the really chancy stuff should go elsewhere," Norman Packard says. "Personally, if I were running the Institute, I would tend to take more chances."

In the spring of 1987, though, the Institute baffled all of its critics. It hired Ed Witten as a permanent professor.

An aged professor of mathematics sits in his darkened room. The day outside the windows is pale and gray and there's not much light shining through, but the professor has no lights on except for a small fluorescent lamp on his desk. The lamp throws a pool of white light across some journal articles and a letter that he's writing. He's been at the Institute now for more than thirty years, and for at least some of the time he's felt a little out of touch with the best new mathematicians, despite the fact that they're just outside his door. He has a remedy for this, however.

"One weakness of the Institute is that it does not produce Ph.D.s, although we have the right to do that," he says. "It has not been a popular idea. I think that is a mistake. I would think that it is not a good idea to separate the permanent members of the Institute from contact with the younger generation. We have members coming for one or two years, that is different. I would like to have them coming at the age when they are not all fixed on something . . . on a particular problem, and staying here a couple of years and writing their thesis. So they can get advice here, so that they can ask the professors more questions. Both parties would benefit from more contact, both the students and the professors."

Although he's a theoretical mathematician in the grand style, divorced from applications, divorced from things, he wishes there were more hustle and bustle at the Institute, that the place were a little more worldly than it is. "Several professors wanted modern biologists to have professorships here," he says. "You would need laboratories, though, and the idea was that this should be an Institute where money was not spent on those things."

Another old-timer—he's been here from the start of the Oppenheimer days—blames the administration for the Institute's fall from glory. "The administration has gotten too top-heavy," he says in his large and silent office. "When Oppenheimer was the director he didn't spend all his time running the place. He used to do physics half the time. He used to have one secretary, a business manager, and a lady who did the housing arrangements. Now there's a director, and an associate director, and each of them has assistants plus secretaries, and sometimes two secretaries. This is poisonous. After a while the administration lives for itself, and the faculty becomes a side-issue."

The administration at the Institute for Advanced Study is curator of icons and protector of . . . *The Image*. The Institute, after all, is the One True Platonic Heaven on earth and it must always look the part. The outside world must know only that the Institute is this austere place where Einstein worked, where everything is always harmonious and serene. If there are a few wrinkles in the fabric of perfection from time to time, these must be smoothed out immediately, or at any rate hidden from view. In no case can they be publicly acknowledged. *None of it must be allowed to get out.*

In the early 1960s, Oppenheimer thought that, in view of the historical importance of the place to the development of science, and more generally to the history of scholarship in the twentieth century, it would be a good idea if an account of the Institute's early years were written up for posterity. And so he commissioned one Beatrice M. Stern, of San Francisco, to research and write the official history of the Institute for Advanced Study.

Mrs. Stern came to Princeton where she was given access to the Institute's files and papers, including minutes of faculty meetings, letters, memoranda, and so on. She interviewed faculty members, temporary members, and just about anyone else who would talk with her. In the end, she worked on the project for a total of nine years and produced a document that runs almost to 800 pages, which she submitted to the Institute's authorities. It then became a disowned and forbidden manuscript, as if it described the slaughter of young virgins in the basement of Fuld Hall or rampant cannibalism among the faculty. Whatever it contained, it was clear that *none of it must be allowed to get out*.

"There is no official Institute history," says Institute director Harry Woolf. "The manuscript by Mrs. Stern is not an official history, and I don't know where a copy of it can be found."

"Why does it seem to be a suppressed document?" I ask him.

"It's not a suppressed document," Woolf says. "It's just that copies of it are so rare."

Someone suggests that Harold Cherniss may have one of these rare copies. "The manuscript is private," Cherniss says. "You can't see it." I ask him if I could get a copy from someone else. "I hope you can't," he says.

Someone suggests that Herman Goldstine may have a copy. "No, I don't have a copy," Goldstine says. "It was a trifling matter anyway. The lady didn't have any training in historical research. It's the story of twenty-six faculty members stabbing each other in the back with knives."

There's a rumor that Deane Montgomery has a copy. "I don't have a copy," Deane Montgomery says. "I was never allowed to have a copy."

John Bahcall has a copy, but he's not letting anyone see it. "I can't let you see the manuscript," he says. "It's a matter for the administration to decide whether it can be given out. I can't give it out."

"Why not?" I ask him. "What's the big problem with it?"

"Well, it's not very well written," Bahcall says.

Beatrice M. Stern, neé Beatrice Mark, graduated from the University of California with a degree in economics in 1918, then went to work for the State of California Compensation Insurance Board. In the 1920s, she married Max Stern, a newspaper reporter from San Francisco. They moved to Washington, D.C., for a while but later came back to California. In the 1950s, Mrs. Stern was hired by the United Parcel Service to write the company history, a project she worked on for about two years.

When she came to the Institute, Beatrice Stern put an inordinate amount of work into her research and writing, finally producing a two-volume manuscript entitled *History of the Institute for Advanced Study 1930–1950*. It is an utterly innocuous document. It's not a record of malfeasance and wrongdoing; it contains no revelations of personal misconduct on the part of anyone living or dead; and by any ordinary and reasonable standard it should cause no embarrassment to the institution whose history it records.

That would be true, anyway, if it were the story on an institution on Planet Earth. As it was, Mrs. Stern apparently forgot that she was chronicling the life of the Platonic Heaven, and so she inadvertently describes, apparently with great fidelity to the records she had access to, the squabbles and disputes that will beset any earthly group of people working in proximity to each other over long periods of time. Her history provides ample evidence that even the finest men and women, even the world's greatest icons of science, are not above pettiness and jealousy, spite and bad judgment. Beatrice Stern's manuscript portrays the Institute's worthies with all their warts, foibles, and imperfections, and this, apparently, is why *none of it must be allowed to get out.*

Later, in the mid-1970s, the whole business was reenacted on a smaller scale. The Institute was gearing up for its Albert Einstein centennial celebration, to be held March 4th through 9th, 1979, and for this occasion it wanted to put together some promotional materials, including a brief historical sketch of the place. To research and write them, associate director John Hunt hired one William G. Wing, who had previously worked for the *New York Herald Tribune.* So Wing started in with his researching and interviewing and writing. He submitted his manuscripts one by one to Hunt, who seemed to like them well enough, and everything was going along swimmingly . . . until Wing got to the Institute's computer project.

The computer project, of course, had been controversial; it had stirred up high tides of emotion at the time, and now Institute people were trying to cover up their tracks, making it hard for Wing to find out who had been for it and who against it, who had invented what part of the computer, who had said what to whom, and so on. It seemed, in fact, that all the old passions were threatening to erupt all over again, and so before matters got entirely out of hand Wing's project was terminated, and that was the end of that.

Although the Institute is not in fact the One True Platonic Heaven, it approaches so closely a state of genuine grace and perfection that one can't help but wonder what it would be like if it were truly the Paradise on Earth it was meant to be. Here, then, the Ideally Perfect, Platonic-heavenly Institute for Advanced Study.

Post-docs, first of all, to remain just as they are. If there's anything the Institute does superbly well, it's in giving itself over to post-doctoral work to the extent that it does. This in fact may be the Institute's greatest contribution to science, for it's from the ranks of the young, by and large—most likely not the Institute regulars, and certainly not the Great Ancients—that important new science is going to come. Here the Institute is already as utopian as it could possibly be.

Harsh reality enters the picture in the form of the permanent faculty. The issues were defined clearly enough in an ancient and famous exchange between J. Robert Oppenheimer and Deane Montgomery:

"I want the best men in the world," Montgomery said.

"I understand that," Oppie said. "But we must consider how they will fit in here harmoniously."

Oppie had it wrong. If there's anything the Institute has too much of already it's concord and placidity. There's no tension on the premises, no crackle in the air, no sense at all that there are mad geniuses lurking about.

"I wish we had more crazy people here," Freeman Dyson has said.

Just so.

The solution, however, is not to bring in graduate students, laboratories, linear accelerators, or any other such artificial aids to confusion. What the Ideally Perfect Institute needs more than anything else are one or two prickly pears on the faculty. Often enough, the world's most creative geniuses are overbearing egotistical snobs, but there are none of these to be found at the Institute, which has managed to attain the benign harmoniousness of a clear blue sky. But it's a fundamental principle of aesthetics that the rarest beauty of all is achieved by the addition of an incongruous detail to an otherwise uniform whole. Violated symmetry, deliberate ambiguity, a little chromatic fantasy amid all the harmonic tones—these make for an organic unity out of what had been a mechanical sameness.

And so it is with institutions. To balance out all the harmony and concordance, the Institute needs some dissonance, some thunder, some strangeness in the proportion . . . .

. . . and it also needs its Great Ancients, its old men shuffling through the hallways, rattling their teacups, and nodding off of an afternoon. In a way, these living icons are the most important people of all: they are the Institute's memory, its link to the past, a vision of Golden Ages far away and long ago. For them the Institute is—and in fact ought to be—a final resting place, a great reward, a toyland.

One morning at about 9:30 Otto Neugebauer walks into the Institute dining hall for breakfast. Neugebauer is literally a man from another era, for he was born in another century, in 1899. He had been at the University of Göttingen back in the halcyon days, back when David Hilbert was director of the Mathematical Institute, when Heisenberg, Pauli, Oppenheimer, Dirac, all the great gods of science, came to Göttingen and founded the new order in physics. The rest of them are all dead now, but here's Otto Neugebauer gliding through the Institute dining hall, crisply dressed in a light brown suit, starched shirt, and tie, looking as if he's ready to take on the board of directors. Watching him cross the room is like seeing a ghost, but there he is, ruddy cheeks and pure white hair, big as life.

Although he's well past retirement age, Neugebauer will go this morning to his office, as he does every single day of the week, for he's still—even after all this time—he's still an active worker. So he'll go up there to the third floor of Fuld Hall and pore over some of his books and papers on ancient astronomy, his favorite subject. He might put in a call to some other expert at the Institute, or on the Princeton faculty, to clarify a point of astrophysics. He might write a few crabbed lines. At 12 o'clock he'll go to lunch, and later on he'll be seen again at tea. But otherwise he's up in his office, reading and working and keeping the flame alive. There's just no stopping Otto Neugebauer.

Neugebauer's finished his breakfast now, and he's just about ready to leave for his office, when Deane Montgomery walks in. Montgomery's a comparative youngster, only seventy-seven, but he's been in his office since 7:30 this morning, reading mathematics and thinking about his contorted topological shapes. He's a living icon, too, a whisper from the ages past. When he stops and chats with Neugebauer it's a metaphysical moment, a bit of history being played out there in front of you.

After a few seconds of old-boy banter, Neugebauer shuffles off to work, and Montgomery goes into the serving area for a glass of tomato juice, which he polishes off in a single draft. On the way out, he drops 50¢ into the money basket. Then both Otto Neugebauer and Deane Montgomery are gone, and the dining hall is empty for a while, empty of its Sons of the Golden Age.

Appendix

Programs for the Mandelbrot Set and Cellular Automata

Two chapters of this book (Chapter 4, "Behold the Forms," and Chapter 10, "Nature's Own Software") discuss mathematical objects that are relatively easy to produce on a personal computer. Below are BASIC programs for the Mandelbrot set and for cellular automata, written by John Milnor and Nicholas Tufillaro, respectively. I acknowledge with thanks their permission to publish these programs here.

```
0 REM: BASIC program for the Mandelbrot set. Program written by
1 REM: John Milnor, Institute for Advanced Study, January 14, 1986.
2 REM: Macintosh version (Microsoft BASIC 2.0) by Linda Eshleman
3 REM: and Ed Regis for monochrome 512 x 342-pixel screen.
4 REM: For more accuracy (but less speed), insert 15 DEFDBL A, B, W-Y
5 REM: Inputs for: a-center     b-center   screenwidth   h0   h1   h2
6 REM: Whole set:   -.765        .21203        2.5       12   18   60
7 REM: Top part:    -.11        1.02            .27      13   18  100
8 REM: Right side:   .442        .338           .06      20   32  120
9 REM: Pix-size: Choose 1 for high-res and slow; 8 for low-res and fast.
10 DEFINT C-P
20 PRINT "Plot of those a +ib for which the orbit of zero under the map"
25 PRINT " x +iy → (x +iy)*(x +iy)  + (a +ib) is bounded"
30 INPUT "a-center"; ACENT: INPUT"b-center"; BCENT: INPUT "screenwidth"; WID
35 PRINT "Color cutoff points (say 10, 20, 60)"
40 INPUT "h0="; H0: INPUT "<h1="; H1: INPUT "<h2="; H2
50 INPUT "pix-size (1,2,4,5 or 8)"; PS
60 ASCALE = PS*WID/512: BSCALE = 1.18*ASCALE
70 AMIN = ACENT -256*ASCALE/PS: BMAX = BCENT +171*BSCALE/PS
80 CLS: REM: Screen now in graphic mode, ready to go
```

```
100 FOR I=0 TO 512/PS-1: A=AMIN  + I*ASCALE
110  FOR J=0 TO 342/PS-1: B=BMAX - J*BSCALE: X=0: Y=0
120   FOR H=0 TO H2
130     XNEW = X*X - Y*Y  + A: REM: Computing (X +iY)*(X +iY) + (A +iB)
140     Y=2*X*Y +B: X=XNEW
150      IF X*X+Y*Y > 342 THEN GOSUB 800 ELSE NEXT H
200 NEXT J,I
300 IF INKEY$=" " THEN 300: REM: Done; waiting for input before restart.
310 PRINT "Center=("; ACENT; BCENT; "), width="; WID
320 PRINT "h=("; H0; H1; H2; ")": GOTO 30
800 IF H<H0 THEN COL=1 ELSE IF H<H1 THEN COL=3 ELSE COL=2
810 FOR K=0 TO PS-1: FOR L=0 TO PS-1: PSET (I*PS +K, J*PS +L), COL:
    NEXT L, K
820 RETURN
```

```
10 REM Cellular Automata Program
15 REM Mod 2 Cellular Automata Rule. Initial value: single site
20 REM Macintosh Microsoft BASIC 2.1 Version
30 REM Copyright 1986 by Nicholas B. Tufillaro.
40 DEFINT i,j,k
50 REM Site arrays, R is current row, Q is previous row
60 DIM R(512), Q(512)
70 DEFINT XMAX, YMAX: XMAX = 512: YMAX = 342
75 REM Initial Value single site at top center of screen.
80 Q(255) = 1: PSET (255, 0)
90 FOR n=1 TO YMAX-1
100   FOR j=0 TO XMAX-1
120     i=j-1
130     IF i=-1 THEN i=XMAX-1
140     k=j +1
150     IF k=XMAX THEN k=0
160     REM Cellular Automata Rule "Mod 2"
170     site=(Q(i)+Q(k)) MOD 2
180     R(j)=site
190     IF site=1 THEN PSET (j,n)
200   NEXT j
210   FOR j=0 TO XMAX-1
220     Q(j)=R(j)
230   NEXT j
240 NEXT n
250 END
```

Sources

General Sources on the Institute for Advanced Study

A Community of Scholars. Institute for Advanced Study. Faculty and Members 1930–1980. Princeton: Institute for Advanced Study, 1980.

Annual Report. Institute for Advanced Study. 1980/81–1984/85.

Bulletin. Nos. 1–12. Institute for Advanced Study, 1930–1946. A series of pamphlets describing the Institute, and listing its staff and members. Published annually except for 1931, 1932, 1942, 1943, and 1944.

Christy, Duncan. "Life in the 'Intellectual Zoo.'" *M* (July 1986): 78.

Corry, John. "Visit to an Intellectual Hotel." *New York Times Magazine* (May 15, 1966): 50.

Davies, John. "The Institute for Advanced Study." *Princeton Magazine* (August, September, October 1982).

Jones, Landon Y., Jr. "Bad Days on Mount Olympus." *The Atlantic* (February 1974): 37. Focuses on the battle to remove director Carl Kaysen.

Kaysen, Carl. *Report of the Director 1966–1976*. Princeton: Institute for Advanced Study, 1976.

Regis, Edward, Jr. "Einstein's Sanctum." *Omni* (September 1984): 88.

Stern, Beatrice M. *A History of the Institute for Advanced Study 1930–1950*. Unpublished 2-volume manuscript, 1964. A microfilm copy of the manuscript is part of the J. Robert Oppenheimer Papers of the Library of Congress, Washington, D.C. (accession file #16646). A carbon copy of the original manuscript is available in the Special Collections of the Hoover Library at Western Maryland College, Westminster, Maryland, 21157.

Stuckey, William K. "The Garden of Lonely Wise." *Science Digest* (February 1975): 28.

Chapter 1. The Platonic Heaven

Blanshard, Frances. *Frank Aydelotte of Swarthmore*. Middletown, Conn.: Wesleyan University Press, 1970. Biography of the Institute's second director.

Flexner, Abraham. *An Autobiography*. New York: Simon and Schuster, 1960. A revision of Flexner's *I Remember*, published in 1940.

Howie, Diana M. "The Legacy of Louis Bamberger." *New Jersey Monthly* (September 1984): 44.

"L. Bamberger, Philanthropist." *Newark Evening News*. March 12, 1944.

"Louis Bamberger." *Sunday Star Ledger*. March 12, 1944.

Chapter 2. The Pope of Physics

Clark, Ronald W. *Einstein: The Life and Times*. New York and Cleveland: World, 1971. A popular account of Einstein's life.

Dukas, Helen, and Banesh Hoffmann (eds.). *Albert Einstein: The Human Side. New Glimpses from His Archives*. Princeton: Princeton University Press, 1979. A collection of Einstein's epigrams and words of wisdom on various subjects, compiled by Einstein's secretary, and by his biographer.

French, A.P. (ed.). *Einstein: A Centenary Volume*. Cambridge, Mass.: Harvard University Press, 1979. Reminiscences of and essays about Einstein; includes selections from Einstein's letters and published works.

Herbert, Nick. *Quantum Reality: Beyond the New Physics*. New York: Doubleday/Anchor, 1985. A popular exposition of the EPR paradox and Bell's theorem.

Hoffmann, Banesh, and Helen Dukas. *Albert Einstein: Creator and Rebel*. New York: Viking, 1972.

Holton, Gerald, and Yehuda Elkana (eds.). *Albert Einstein: Historical and Cultural Perspectives*. Princeton: Princeton University Press, 1982.

Maranto, Gina. "Einstein's Brain." *Discover* (May 1985): 29. The strange saga of Albert Einstein's sole earthly remains.

Mermin, N. David. "Is the Moon There When Nobody Looks? Reality and the Quantum Theory." *Physics Today* (April 1985). The EPR paradox, Bell's theorem, and "spooky actions at a distance."

Pais, Abraham. *'Subtle is the Lord . . . ' The Science and the Life of Albert Einstein*. Oxford and New York: Oxford University Press, 1982. An authoritative account of

Einstein's life and science by a physicist who knew Einstein during his years at the Institute.

Sayen, Jamie. *Einstein in America.* New York: Crown Publishers, 1985.

Schilpp, Paul Arthur (ed.). *Albert Einstein: Philosopher-Scientist.* 2 vols. 1949. LaSalle, Ill.: Open Court, 1982. Important 2-volume collection of papers about Einstein; contains Einstein's autobiography, and a reply to his critics.

Wheeler, John Archibald, and Wojciech Zurek (eds.). *Quantum Theory and Measurement.* Princeton: Princeton University Press, 1983. A collection of papers on the EPR paradox and on Bell's inequality; includes the original paper by Einstein, Poldosky, and Rosen, as well as Niels Bohr's replies.

Woolf, Harry (ed.). *Some Strangeness in the Proportion.* Reading, Mass.: Addison-Wesley, 1980. Proceedings of a centennial symposium in honor of Albert Einstein held at the Institute for Advanced Study in March, 1979.

Chapter 3. The Grand High Exalted Mystical Ruler

Benacerraf, Paul, and Hilary Putnam (eds.). *Philosophy of Mathematics: Selected Readings.* Englewood Cliffs, N.J.: Prentice-Hall, 1964. Contains Gödel's papers "Russell's Mathematical Logic" and "What Is Cantor's Continuum Problem?"

Dawson, John (trans. and ed.). "Discussion on the Foundation of Mathematics." *History and Philosophy of Logic* 5 (1984): 111. An English translation of the discussion following Kurt Gödel's first public announcement of his incompleteness theorem. Remarks by Hans Hahn, Rudolf Carnap, John von Neumann, Arnold Scholz, Arend Heyting, and Kurt Gödel.

______. "Kurt Gödel in Sharper Focus." *The Mathematical Intelligencer* 6 (1984): 9. A short biographical sketch of Kurt Gödel; includes photographs of Gödel's childhood home in Brno, Czechoslovakia, a page from Gödel's first arithmetic workbook, and Gödel's report card for February, 1917, showing that he received the highest possible grade for every subject, except for mathematics.

______. "Cataloguing the Gödel *Nachlass* at the Institute for Advanced Study." *Abstracts of the 7th International Congress of Logic, Methodology, and Philosophy of Science* 6 (1983): 59.

Feferman, Solomon, et al. (eds.). *Kurt Gödel: Collected Works.* Volume 1: Publications 1929–1936. Oxford: Oxford University Press, 1986. Includes an important biographical essay by Feferman, one of the few sources of information about the life of Kurt Gödel.

Gödel, Kurt. "An Example of a New Type of Cosmological Solutions of Einstein's Field Equations of Gravitation." *Reviews of Modern Physics* 21 (July 1949): 447. Presents Gödel's theory of time travel.

_____. "A Remark about the Relationship between Relativity Theory and Idealistic Philosophy." In Schilpp volume, above.

Heijenoort, Jean van (ed.). *From Frege to Gödel*. Cambridge, Mass.: Harvard University Press, 1967. Contains an English translation of Gödel's famous 1931 paper on undecidable propositions.

Hofstadter, Douglas R. *Gödel, Escher, Bach: An Eternal Golden Braid*. New York: Vintage Books, 1980. An original, profound, and brilliant book; by any standard an incomparable reading experience.

Kline, Morris. *Mathematics: The Loss of Certainty*. New York: Oxford University Press, 1980. Discusses Gödel's results in a chapter entitled "Disasters."

Kreisel, G. "Kurt Gödel." *Biographical Memoirs of Fellows of the Royal Society*. 26 (1981): 148. An exhaustive, sometimes witty survey of Gödel's life and work; contains some frank discussion of Gödel's personality and depressions, by someone who knew him during the Princeton years.

Rucker, Rudy. *Infinity and the Mind*. New York: Bantam Books, 1982. An attempt to convince the reader that infinity is real; comes "complete with illustrations, puzzles and paradoxes to solve," and includes the author's "Conversations with Gödel."

Chapter 4. Behold the Forms

Albers, Donald J., and G. L. Alexanderson (eds.). *Mathematical People: Profiles and Interviews*. Boston: Birkhäuser, 1985.

Bourbaki, Nicholas [André Weil?]. "The Architecture of Mathematics." *American Mathematical Monthly* 57 (1950): 221.

Campbell, Douglas M., and John C. Higgins (eds.). *Mathematics: People, Problems, Results*. 3 vols. Belmont, Calif.: Wadsworth, 1984.

Davis, Philip J., and Reuben Hersh. *The Mathematical Experience*. Boston: Houghton Mifflin, 1981.

Dewdney, A. K. "Computer Recreations." *Scientific American* (August 1985): 16. Hints on programming the Mandelbrot set.

Dieudonné, Jean. "The Work of Nicholas Bourbaki." *American Mathematical Monthly* 77 (1970): 134.

Gardner, Martin. "Mathematical Games." *Scientific American* (December 1976): 124. Describes "monster curves," such as the Hilbert curve and the Koch snowflake.

Gleick, James. "The Man Who Reshaped Geometry." *New York Times Magazine* (December 8, 1985): 64. A profile of Benoit Mandelbrot.

Grillo, John P. "Fractal Trees." *Nibble Mac* (January/February, 1986): 48. Gives the program (written in Microsoft BASIC for the Macintosh) for producing the fractal tree pictured on page 91 of the text.

Halmos, Paul. "Nicolas Bourbaki." *Scientific American* (May 1957): 88.

Henney, Dagmar Renate. "Bourbaki. A French General—or a Mysterious Society?" *Mathematics Magazine* 36 (September-October 1963): 252.

Kline, Morris. *Mathematics and the Search for Knowledge*. New York: Oxford University Press, 1985.

Mandelbrot, Benoit B. Interview. *Omni* (February 1984): 65.

——. *The Fractal Geometry of Nature* (rev. ed.), New York: W.H. Freeman, 1983.

Newman, James R. (ed.). *The World of Mathematics*. 4 vols. New York: Simon and Schuster, 1956.

Reid, Constance. *Hilbert*. New York: Springer-Verlag, 1970.

Richards, Ian. "Number Theory." *Mathematics Today* (ed., Lyn Arthur Steen). New York: Springer-Verlag, 1978.

Stein, Kathleen. "The Fractal Cosmos." *Omni* (February 1983): 63.

Weil, André. "The Future of Mathematics." *American Mathematical Monthly* 57 (1959): 295.

Chapter 5. Good Time Johnny

Bass, Thomas. *The Eudaemonic Pie*. Boston: Houghton Mifflin, 1985. The early adventures of J. Doyne Farmer, Norman Packard, and friends as they attempt to beat roulette with dynamical systems theory. Excerpts appeared in *Science Digest*, April and May, 1985.

Blair, Clay. "Passing of a Great Mind." *Life* (February 25, 1957): 89. Profile of John von Neumann, published shortly after his death.

Dyson, Freeman. *Disturbing the Universe*. New York: Harper & Row, 1979. Chapter 18, "Thought Experiments," is a recollection of John von Neumann and his work at the Institute.

_____. "The Future of Physics." *Physics Today* (1970): 25. Contains Dyson's evalualuations of the Institute for Advanced Study computer project, and of the Institute itself.

Goldstine, Herman H. *The Computer from Pascal to von Neumann.* Princeton: Princeton University Press, 1980. Contains several chapters on von Neumann and the IAS computer; written by an Institute member who worked on the ENIAC as well as on the IAS computer project.

Goldstine, Herman H. , and Eugene Wigner. "Scientific Work of J. von Neumann." *Science* (1957): 683.

Halmos, P. R. "The Legend of John von Neumann." *American Mathematical Monthly* 80 (1973): 382. Reprinted in Douglas M. Campbell and John C. Higgins (eds.)., *Mathematics: People, Problems, Results.* 3 vols. Belmont, Calif.: Wadsworth, 1984.

Heims, Steve J. *John von Neumann and Norbert Wiener.* Cambridge, Mass.: The MIT Press, 1980. A dual biography.

Kemeny, John G. "Man Viewed as a Machine." *Scientific American* 192 (1955): 58. An elementary account of the Turing machine and von Neumann's self-reproducing automata.

Tipler, Frank J. "Extraterrestrial Intelligent Beings Do Not Exist." *Quarterly Journal of the Royal Astronomical Society* 21 (1980): 267. Reprinted in Edward Regis, Jr. (ed.)., *Extraterrestrials: Science and Alien Intelligence.* Cambridge, England: Cambridge University Press, 1985.

Ulam, S. M. *Adventures of a Mathematician.* New York: Scribner's, 1983. Includes many recollections of von Neumann by a close friend and fellow mathematician.

_____. "John von Neumann 1903–1957." *Bulletin of the American Mathematical Society* 64 (1958): 1. Extensive survey of von Neumann's life and work.

von Neumann, John. *Theory of Self-Reproducing Automata* (edited and completed by Arthur Burks). Urbana and London: University of Illinois Press, 1966.

Wiener, Norbert. *I Am a Mathematician.* Cambridge, Mass.: The MIT Press, 1956. Includes reminiscences of von Neumann and Julian Bigelow.

Chapter 6. The Nim-Nim-Nim Man

Bacher, Robert F. "Robert Oppenheimer." *Proceedings of the American Philosophical Society* 116 (1972): 279.

Bernstein, Barton J. "In the Matter of J. Robert Oppenheimer." *Historical Studies in the Physical Sciences* 12 (1982): 195. Detailed account of the Oppenheimer hearings; makes use of recently declassified FBI reports and other "secret" information.

———. "The Oppenheimer Conspiracy." *Discover* (March 1985): 22. A popular version of the above.

Bernstein, Jeremy. "A Question of Parity." *A Comprehensible World*. New York: Random House, 1967. The lives and work of T.D. Lee and C.N. Yang.

Goodchild, Peter. *J. Robert Oppenheimer: Shatterer of Worlds*. Boston: Houghton Mifflin, 1981. Profusely illustrated biography.

Hammond, Lansing V. "A Meeting with Robert Oppenheimer." Unpublished manuscript, 1979.

Kipphardt, Heinar. *In the Matter of J. Robert Oppenheimer* (trans. Ruth Speirs). New York: Hill and Wang, 1968. A play based on the Oppenheimer hearing transcripts; Oppie's reaction was that the author tried to convert a farce into a tragedy.

Lamont, Lansing. *Day of Trinity*. New York: Atheneum, 1965. A vivid and dramatic portrayal of the Los Alamos scientists and the Trinity test.

Lee, T.D. "Broken Parity." *T.D. Lee: Selected Papers*. Volume 3. Boston: Birkhäuser, 1986. Section 4, "Broken Friendship," tells Lee's version of his break with Frank Yang.

Michelmore, Peter. *The Swift Years: The Robert Oppenheimer Story*. New York: Dodd, Mead, 1969.

Morrison, Philip. "The Overthrow of Parity." *Scientific American* 196 (1957): 45. The Yang and Lee work on the breakdown of mirror symmetry.

"A New World, A Mystic World." *Time* 126 (July 29, 1985): 40. Los Alamos and Trinity revisited, forty years after the event.

Oppenheimer, J. Robert. Interview by Thomas S. Kuhn. *American Institute of Physics* (November 18 and 20, 1963).

Oppenheimer, J.R., and Robert Serber. "On the Stability of Stellar Neutron Cores." *Physical Review* 54 (October 1, 1938): 540.

Oppenheimer, J.R., and H. Snyder. "On Continued Gravitational Contraction." *Physical Review* 56 (September 1, 1939): 455.

Oppenheimer, J.R., and G.M. Volkoff. "On Massive Neutron Cores." *Physical Review* 55 (February 15, 1939): 374.

Rabi, I.I., et al. *Oppenheimer*. New York: Scribner's, 1969. Recollections and reminiscences by I.I. Rabi, Robert Serber, Victor Weisskopf, Abraham Pais, and Glenn Seaborg.

Smith, Alice Kimball, and Charles Weiner (eds.). *Robert Oppenheimer: Letters and Recollections*. Cambridge, Mass.: Harvard University Press, 1980. An important reference work for Oppenheimer's life through 1945, spotty thereafter.

Strauss, Lewis L. *Men and Decisions*. New York: Doubleday, 1962.

Szasz, Ferenc Morton. *The Day the Sun Rose Twice: The Story of the Trinity Site Nuclear Explosion July 16, 1945*. Albuquerque: University of New Mexico Press, 1984. Corrects some errors in Lansing Lamont's book on the subject.

United States Atomic Energy Commission. *In the Matter of J. Robert Oppenheimer: Transcript of Hearing before Personnel Security Board, Washington, D.C., April 12, 1954 through May 6, 1954*. Washington, D.C.: Government Printing Office, 1954.

Wilson, Jane (ed.). *All in Our Time: The Reminiscences of Twelve Nuclear Pioneers*. Chicago: Bulletin of the Atomic Scientists, 1975.

Yang, Chen Ning. *Selected Papers 1945–1980 with Commentary*. San Francisco: W.H. Freeman, 1983. Contains Yang's version of his break with T.D. Lee.

Chapter 7. Hubble, Bubble Toil and Trouble

Bartusiak, Marcia. "The Bubbling Universe." *Science Digest* (February 1986): 64.

Davis, Marc, Piet Hut, and Richard A. Mueller. "Extinction of Species by Periodic Comet Showers." *Nature* 308 (April 19, 1984): 715.

de Lapparent, Valérie, Margaret J. Geller, and John P. Huchra. "A Slice of the Universe." Harvard-Smithsonian Center for Astrophysics. Preprint 2231 (1985). *Astrophysical Journal (Letters)* 302 (1986).

Kutner, M.L., et al. "The Molecular Complexes in Orion." *Astrophysical Journal* 215 (July 15, 1977): 521.

Schwarzschild, Bertram. "Redshift Surveys of Galaxies Find a Bubbly Universe." *Physics Today* (May 1986): 17.

Chapter 8. Carrying the Fire

[Adler, Stephen L.] "Physics and Astronomy at the Institute for Advanced Study, 1960–1980." Unpublished manuscript, n.d.

Bernstein, Jeremy. "Pauli's Puzzle." *Science Digest* (August 1986): 41. Wolfgang Pauli and the neutrino.

Crease, Robert P., and Charles C. Mann. *The Second Creation: Makers of the Revolution in Twentieth-Century Physics*. New York: Macmillan, 1986. A brilliantly written history of particle physics. Emphasizes personalities, but succeeds in explaining complex theories clearly and vividly.

Feynman, Richard P. *QED: The Strange Theory of Light and Matter*. Princeton: Princeton University Press, 1985. A clear and witty account of what is *known* in physics, as contrasted with what is still a matter of research and speculation.

Feynman, Richard P., et al. *The Feynman Lectures on Physics*. 3 vols. Reading, Mass.: Addison-Wesley, 1963. *Time* magazine once asked Carl Sagan to list the six books that any educated person ought to have read; *The Feynman Lectures*, volume one, was high on Sagan's list.

Pais, Abraham. *Inward Bound: Of Matter and Forces in the Physical World*. Oxford: Clarendon Press, 1986. A history of particle physics that emphasizes science rather than people, written by one of the participants who seemed to be on a first-name basis with everyone else in the discipline.

Chapter 9. The Truth About Things

Bahcall, John. "The Problem of Solar Neutrinos." Vassar Seminar on the Frontiers in the Natural Sciences, Vassar College, Poughkeepsie, New York. Unpublished draft transcript, February 26, 1976.

Bahcall, John N., and Raymond Davis, Jr. "An Account of the Development of the Solar Neutrino Problem." Charles A. Barnes et al. (eds.)., *Essays in Nuclear Astrophysics*. Cambridge, England: Cambridge University Press, 1982.

______. "Solar Neutrinos: A Scientific Puzzle." *Science* 191 (January 23, 1976): 264.

Finkbiner, Ann. "Paradigm Lost?" *Johns Hopkins Magazine* (June 1985): 25. Account of a talk given by Kuhn at Johns Hopkins University; scientists polled voted ten to one that they were describing "objective reality" and not a mere "interpretation."

Kuhn, Thomas S. *The Essential Tension*. Chicago: University of Chicago Press, 1977.

______. *The Structure of Scientific Revolutions*, 2nd ed. Chicago: University of Chicago Press, 1970.

Pollie, Robert. "Brother, Can You Paradigm?" *Science 83* (July/August 1983): 76. One of the greatest titles in all magazinedom; adventures in paradigm chic and cosmic one-upmanship.

Shapere, Dudley. "External and Internal Factors in the Development of Science." *Science and Technology Studies* 4 (1986): 1.

______. "The Concept of Observation in Science and Philosophy." *Philosophy of Science* 49 (1982): 485.

———. *Reason and the Search for Knowledge*. Dordrecht, The Netherlands: Reidel, 1983.

———. "The Paradigm Concept." *Science* 172 (May 14, 1971): 706. Review of Kuhn's book *The Structure of Scientific Revolutions*, by his most persistent critic.

Chapter 10. Nature's Own Software

Dewdney, A.K. "Computer Recreations." *Scientific American* 252 (May 1985): 18. Contains Stephen Wolfram's challenge to hackerdom.

Levy, Steven. "The Portable Universe." *Whole Earth Review* (Winter 1985): 42. Stephen Wolfram and cellular automata, by the author of *Hackers*.

Poundstone, William. *The Recursive Universe: Cosmic Complexity and the Limits of Scientific Knowledge*. New York: Morrow, 1985. A detailed description of the game of Life; also discusses von Neumann's self-reproducing automata.

Wolfram, Stephen. "Cellular Automata." *Los Alamos Science* (Fall 1983): 2. Good introductory-level account of cellular automata, with many illustrations, including one of Wolfram holding his famous seashell.

———. "Cellular Automata as Models of Complexity." *Nature* 311 (October 4, 1984): 419. More technical account.

———. "Computer Software in Science and Mathematics." *Scientific American* 251(September 1984): 188. A clear but profound exposition of how cellular automata may provide unique models for the study of irreducibly complex phenomena.

———. "Statistical Mechanics of Cellular Automata." *Reviews of Modern Physics* 55 (1983): 601. Extensive technical discussion and classification of cellular automata and their applications to problems in physics.

Chapter 11. Beyond the Invisible

Crease, Robert P., and Charles C. Mann. "The Gospel of String." *The Atlantic* (April 1986): 24. String theory for beginners.

Freedman, Daniel Z., and Peter van Nieuwenhuizen. "The Hidden Dimensions of Spacetime." *Scientific American* (March 1985): 74. How the Kaluza-Klein maneuver hides unwanted dimensions.

Ginsparg, Paul, and Sheldon Glashow. "Desperately Seeking Superstrings?" *Physics Today* (May 1986): 7. The authors answer the question, "Why is the smart money all tied up in strings?"

Green, Michael B. "Superstrings." *Scientific American* 255 (September 1986): 48. An authoritative introduction to the superstring universe.

______. "Unification of Forces and Particles in Superstring Theories." *Nature* 314 (April 4, 1985): 409. A semitechnical presentation of superstring theory.

Kolb, Edward W., et al. "The Shadow World of Superstring Theories." *Nature* 314 (April 4, 1985): 415. "The reader could be living in the middle of a shadow mountain or at the bottom of a shadow ocean," the authors say.

Schwarz, John H. "Completing Einstein." *Science 85* (November 1985): 60. An elementary exposition of superstring theory by the person most responsible for its success.

______. "Dual-Resonance Models of Elementary Particles." *Scientific American* (February 1975): 61. Regge trajectories, Veneziano amplitudes, and the "old" string theory.

Schwarzschild, Bertram M. "Anomaly Cancellation Launches Superstring Bandwagon." *Physics Today* (July 1985): 17. A review of recent progress in string theory.

Taubes, Gary. "Everything's Now Tied to Strings." *Discover* (November 1986): 34. A popular account.

Chapter 12. Babes in Toyland

Feynman, Richard P. *"Surely You're Joking, Mr. Feynman!"* New York: Norton, 1985. Not a dull page in the book.

Index